Tucholsky Wagner Zola Scott Sydow Freud Schlegel
Turgenev Wallace Fonatne
Twain Walther von der Vogelweide Fouqué Friedrich II. von Preußen
Weber Freiligrath Frey
Fechner Weiße Rose von Fallersleben Kant Ernst Frommel
Fichte Richthofen
Engels Fielding Hölderlin Tacitus Dumas
Fehrs Faber Flaubert Eichendorff
Feuerbach Maximilian I. von Habsburg Fock Eliasberg Zweig Ebner Eschenbach
Ewald Eliot Vergil
Goethe Elisabeth von Österreich London
Mendelssohn Balzac Shakespeare Dostojewski Ganghofer
Trackl Lichtenberg Rathenau Doyle Gjellerup
Mommsen Stevenson Tolstoi Hambruch Droste-Hülshoff
Thoma Lenz Hanrieder
Dach von Arnim Hägele Hauff Humboldt
Reuter Verne Hagen Hauptmann Gautier
Karrillon Garschin Rousseau Baudelaire
Defoe Hebbel
Damaschke Descartes Hegel Kussmaul Herder
Wolfram von Eschenbach Dickens Schopenhauer Rilke George
Bronner Darwin Melville Grimm Jerome Bebel Proust
Campe Horváth Aristoteles Voltaire Federer Herodot
Bismarck Vigny Barlach Heine
Gengenbach
Storm Casanova Tersteegen Grillparzer Georgy
Chamberlain Lessing Langbein Gilm Gryphius
Brentano Lafontaine
Strachwitz Claudius Schiller Kralik Iffland Sokrates
Katharina II. von Rußland Bellamy Schilling
Gerstäcker Raabe Gibbon Tschechow
Löns Hesse Hoffmann Gogol Wilde Vulpius
Luther Heym Hofmannsthal Gleim
Roth Klee Hölty Morgenstern Goedicke
Luxemburg Heyse Klopstock Kleist
Machiavelli La Roche Puschkin Homer Mörike Musil
Navarra Aurel Musset Horaz
Nestroy Marie de France Kierkegaard Kraft Kraus
Lamprecht Kind Kirchhoff Hugo Moltke
Laotse Ipsen Liebknecht
Nietzsche Nansen Ringelnatz
Marx Lassalle Gorki Klett Leibniz
von Ossietzky May
vom Stein Lawrence Irving
Petalozzi Knigge
Platon Kafka
Sachs Pückler Michelangelo Kock
Poe Liebermann Korolenko
de Sade Praetorius Mistral Zetkin

The publishing house tredition has created the series **TREDITION CLASSICS**. It contains classical literature works from over two thousand years. Most of these titles have been out of print and off the bookstore shelves for decades.

The book series is intended to preserve the cultural legacy and to promote the timeless works of classical literature. As a reader of a **TREDITION CLASSICS** book, the reader supports the mission to save many of the amazing works of world literature from oblivion.

The symbol of **TREDITION CLASSICS** is Johannes Gutenberg (1400 – 1468), the inventor of movable type printing.

With the series, tredition intends to make thousands of international literature classics available in printed format again – worldwide.

All books are available at book retailers worldwide in paperback and in hardcover. For more information please visit: www.tredition.com

tredition was established in 2006 by Sandra Latusseck and Soenke Schulz. Based in Hamburg, Germany, tredition offers publishing solutions to authors and publishing houses, combined with worldwide distribution of printed and digital book content. tredition is uniquely positioned to enable authors and publishing houses to create books on their own terms and without conventional manufacturing risks.

For more information please visit: www.tredition.com

The Number Concept Its Origin and Development

Levi Leonard Conant

Imprint

This book is part of the TREDITION CLASSICS series.

Author: Levi Leonard Conant
Cover design: toepferschumann, Berlin (Germany)

Publisher: tredition GmbH, Hamburg (Germany)
ISBN: 978-3-8491-7268-8

www.tredition.com
www.tredition.de

Preface.

In the selection of authorities which have been consulted in the preparation of this work, and to which reference is made in the following pages, great care has been taken. Original sources have been drawn upon in the majority of cases, and nearly all of these are the most recent attainable. Whenever it has not been possible to cite original and recent works, the author has quoted only such as are most standard and trustworthy. In the choice of orthography of proper names and numeral words, the forms have, in almost all cases, been written as they were found, with no attempt to reduce them to a systematic English basis. In many instances this would have been quite impossible; and, even if possible, it would have been altogether unimportant. Hence the forms, whether German, French, Italian, Spanish, or Danish in their transcription, are left unchanged. Diacritical marks are omitted, however, since the proper key could hardly be furnished in a work of this kind.

With the above exceptions, this study will, it is hoped, be found to be quite complete; and as the subject here investigated has never before been treated in any thorough and comprehensive manner, it is hoped that this book may be found helpful. The collections of numeral systems illustrating the use of the binary, the quinary, and other number systems, are, taken together, believed to be the most extensive now existing in any language. Only the cardinal numerals have been considered. The ordinals present no marked peculiarities which would, in a work of this kind, render a separate discussion necessary. Accordingly they have, though with some reluctance, been omitted entirely.

Sincere thanks are due to those who have assisted the author in the preparation of his materials. Especial acknowledgment should be made to Horatio Hale, Dr. D. G. Brinton, Frank Hamilton Cushing, and Dr. A. F. Chamberlain.

Worcester, Mass., Nov. 12, 1895.

Contents.

The Number Concept: Its Origin And Development.

Chapter I.

Counting.

Among the speculative questions which arise in connection with the study of arithmetic from a historical standpoint, the origin of number is one that has provoked much lively discussion, and has led to a great amount of learned research among the primitive and savage languages of the human race. A few simple considerations will, however, show that such research must necessarily leave this question entirely unsettled, and will indicate clearly that it is, from the very nature of things, a question to which no definite and final answer can be given.

Among the barbarous tribes whose languages have been studied, even in a most cursory manner, none have ever been discovered which did not show some familiarity with the number concept. The knowledge thus indicated has often proved to be most limited; not extending beyond the numbers 1 and 2, or 1, 2, and 3. Examples of this poverty of number knowledge are found among the forest tribes of Brazil, the native races of Australia and elsewhere, and they are considered in some detail in the next chapter. At first thought it seems quite inconceivable that any human being should be destitute of the power of counting beyond 2. But such is the case; and in a few instances languages have been found to be absolutely destitute of pure numeral words. The Chiquitos of Bolivia had no real numerals whatever, [1] but expressed their idea for "one" by the word *etama*, meaning alone. The Tacanas of the same country have no numerals except those borrowed from Spanish, or from Aymara or Peno, languages with which they have long been in contact. [2] A few other South American languages are almost equally destitute of numeral words. But even here, rudimentary as the number sense undoubtedly is, it is not wholly lacking; and some indirect expression, or some form of circumlocution, shows a conception of the difference between *one* and *two*, or at least, between *one* and *many*.

These facts must of necessity deter the mathematician from seeking to push his investigation too far back toward the very origin of number. Philosophers have endeavoured to establish certain propositions concerning this subject, but, as might have been expected, have failed to reach any common ground of agreement. Whewell has maintained that "such propositions as that two and three make five are necessary truths, containing in them an element of certainty beyond that which mere experience can give." Mill, on the other hand, argues that any such statement merely expresses a truth derived from early and constant experience; and in this view he is heartily supported by Tylor. [3] But why this question should provoke controversy, it is difficult for the mathematician to understand. Either view would seem to be correct, according to the standpoint from which the question is approached. We know of no language in which the suggestion of number does not appear, and we must admit that the words which give expression to the number sense would be among the early words to be formed in any language. They express ideas which are, at first, wholly concrete, which are of the greatest possible simplicity, and which seem in many ways to be clearly understood, even by the higher orders of the brute creation. The origin of number would in itself, then, appear to lie beyond the proper limits of inquiry; and the primitive conception of number to be fundamental with human thought.

In connection with the assertion that the idea of number seems to be understood by the higher orders of animals, the following brief quotation from a paper by Sir John Lubbock may not be out of place: "Leroy ... mentions a case in which a man was anxious to shoot a crow. 'To deceive this suspicious bird, the plan was hit upon of sending two men to the watch house, one of whom passed on, while the other remained; but the crow counted and kept her distance. The next day three went, and again she perceived that only two retired. In fine, it was found necessary to send five or six men to the watch house to put her out in her calculation. The crow, thinking that this number of men had passed by, lost no time in returning.' From this he inferred that crows could count up to four. Lichtenberg mentions a nightingale which was said to count up to three. Every day he gave it three mealworms, one at a time. When it had finished one it returned for another, but after the third it knew that the feast was over.… There is an amusing and suggestive re-mark in Mr. Galton's interesting *Narrative of an Explorer in Tropical South Africa*. After describing the Demara's weakness in calculations, he says: 'Once while I watched a Demara floundering hopelessly in a calculation on one side of me, I observed, "Dinah," my spaniel, equally embarrassed on the other; she was overlooking half a dozen of her new-born puppies, which had been removed two or three times from her, and her anxiety was excessive, as she tried to find out if they were all present, or if any were still missing. She kept puzzling and running her eyes over them backwards and forwards, but could not satisfy herself. She evidently had a vague notion of counting, but the figure was too large for her brain. Taking the two as they stood, dog and Demara, the comparison reflected no great honour on the man.…' According to my bird-nesting recollections, which I have refreshed by more recent experience, if a nest contains four eggs, one may safely be taken; but if two are removed, the bird generally deserts. Here, then, it would seem as if we had some reason for supposing that there is sufficient intelligence to distinguish three from four. An interesting consideration arises with reference to the number of the victims allotted to each cell by the solitary wasps. One species of Ammophila considers one large caterpillar of *Noctua segetum* enough; one species of Eumenes supplies its young with five victims; another 10, 15, and even up to 24. The number appears to be constant in each species. How does the insect know when her task is fulfilled? Not by the cell being filled, for if some be removed, she does not replace them. When she has brought her complement she considers her task accomplished, whether the victims are still there or not. How, then, does she know when she has made up the number 24? Perhaps it will be said that each species feels some mysterious and innate tendency to provide a certain number of victims. This would, under no circumstances, be any explanation; but it is not in accordance with the facts. In the genus Eumenes the males are much smaller than the females.… If the egg is male, she supplies five; if female, 10 victims. Does she count? Certainly this seems very like a commencement of arithmetic." [4]

Many writers do not agree with the conclusions which Lubbock reaches; maintaining that there is, in all such instances, a perception of greater or less quantity rather than any idea of number. But a careful consideration of the objections offered fails entirely to weaken the argument. Example after example of a nature similar to those just quoted might be given, indicating on the part of animals a perception of the difference between 1 and 2, or between 2 and 3 and 4; and any reasoning which tends to show that it is quantity rather than number which the animal perceives, will apply with equal force to the Demara, the Chiquito, and the Australian. Hence the actual origin of number may safely be excluded from the limits of investigation, and, for the present, be left in the field of pure speculation.

A most inviting field for research is, however, furnished by the primitive methods of counting and of giving visible expression to the idea of number. Our starting-point must, of course, be the sign language, which always precedes intelligible speech; and which is so convenient and so expressive a method of communication that the human family, even in its most highly developed branches, never wholly lays it aside. It may, indeed, be stated as a universal law, that some practical method of numeration has, in the childhood of every nation or tribe, preceded the formation of numeral words.

Practical methods of numeration are many in number and diverse in kind. But the one primitive method of counting which seems to have been almost universal throughout all time is the finger method. It is a matter of common experience and observation that every child, when he begins to count, turns instinctively to his fingers; and, with these convenient aids as counters, tallies off the little number he has in mind. This method is at once so natural and obvious that there can be no doubt that it has always been employed by savage tribes, since the first appearance of the human race in remote antiquity. All research among uncivilized peoples has tended to confirm this view, were confirmation needed of anything so patent. Occasionally some exception to this rule is found; or some variation, such as is presented by the forest tribes of Brazil, who, instead of counting on the fingers themselves, count on the joints of their fingers. [5] As the entire number system of these tribes appears to be limited to *three*, this variation is no cause for surprise.

The variety in practical methods of numeration observed among savage races, and among civilized peoples as well, is so great that any detailed account of them would be almost impossible. In one region we find sticks or splints used; in another, pebbles or shells; in another, simple scratches, or notches cut in a stick, Robinson Crusoe fashion; in another, kernels or little heaps of grain; in another, knots on a string; and so on, in diversity of method almost endless. Such are the devices which have been, and still are, to be found in the daily habit of great numbers of Indian, negro, Mongolian, and Malay tribes; while, to pass at a single step to the other extremity of intellectual development, the German student keeps his beer score by chalk marks on the table or on the wall. But back of all these devices, and forming a common origin to which all may be referred, is the universal finger method; the method with which all begin, and which all find too convenient ever to relinquish entirely, even though their civilization be of the highest type. Any such mode of counting, whether involving the use of the fingers or not, is to be regarded simply as an extraneous aid in the expression or comprehension of an idea which the mind cannot grasp, or cannot retain, without assistance. The German student scores his reckoning with chalk marks because he might otherwise forget; while the Andaman Islander counts on his fingers because he has no other method of counting, — or, in other words, of grasping the idea of number. A single illustration may be given which typifies all practical methods of numeration. More than a century ago travellers in Madagascar observed a curious but simple mode of ascertaining the number of soldiers in an army. [6] Each soldier was made to go through a passage in the presence of the principal chiefs; and as he went through, a pebble was dropped on the ground. This continued until a heap of 10 was obtained, when one was set aside and a new heap begun. Upon the completion of 10 heaps, a pebble was set aside to indicate 100; and so on until the entire army had been numbered. Another illustration, taken from the very antipodes of Madagascar, recently found its way into print in an incidental manner, [7] and is so good that it deserves a place beside de Flacourt's time-honoured example. Mom Cely, a Southern negro of unknown age, finds herself in debt to the storekeeper; and, unwilling to believe that the amount is as great as he represents, she proceeds to investigate the matter in her own peculiar way. She had "kept a tally of these purchases by means of a string, in which she tied commemorative knots." When her creditor "undertook to make the matter clear to Cely's comprehension, he had to proceed upon a system of her own devising. A small notch was cut in a smooth white stick for every dime she owed, and a large notch when the dimes amounted to a dollar; for every five dollars a string was tied in the fifth big notch, Cely keeping tally by the knots in her bit of twine; thus, when two strings were tied about the stick, the ten dollars were seen to be an indisputable fact." This interesting method of computing the amount of her debt, whether an invention of her own or a survival of the African life of her parents, served the old negro woman's purpose perfectly; and it illustrates, as well as a score of examples could, the methods of numeration to which the children of barbarism resort when any number is to be expressed which exceeds the number of counters with which nature has provided them. The fingers are, however, often employed in counting numbers far above the first decade. After giving the Il-Oigob numerals up to 60, Müller adds: [8] "Above 60 all numbers, indicated by the proper figure pantomime, are expressed by means of the word *ipi*." We know, moreover, that many of the American Indian tribes count one ten after another on their fingers; so that, whatever number they are endeavouring to indicate, we need feel no surprise if the savage continues to use his fingers throughout the entire extent of his counts. In rare instances we find tribes which, like the Mairassis of the interior of New Guinea, appear to use nothing but finger pantomime. [9] This tribe, though by no means destitute of the number sense, is said to have no numerals whatever, but to use the single word *awari* with each show of fingers, no matter how few or how many are displayed.

In the methods of finger counting employed by savages a considerable degree of uniformity has been observed. Not only does he use his fingers to assist him in his tally, but he almost always begins with the little finger of his left hand, thence proceeding towards the thumb, which is 5. From this point onward the method varies. Sometimes the second 5 also is told off on the left hand, the same order being observed as in the first 5; but oftener the fingers of the right hand are used, with a reversal of the order previously employed; *i.e.* the thumb denotes 6, the index finger 7, and so on to the little finger, which completes the count to 10.

At first thought there would seem to be no good reason for any marked uniformity of method in finger counting. Observation among children fails to detect any such thing; the child beginning, with almost entire indifference, on the thumb or on the little finger of the left hand. My own observation leads to the conclusion that very young children have a slight, though not decided preference for beginning with the thumb. Experiments in five different primary rooms in the public schools of Worcester, Mass., showed that out of a total of 206 children, 57 began with the little finger and 149 with the thumb. But the fact that nearly three-fourths of the children began with the thumb, and but one-fourth with the little finger, is really far less significant than would appear at first thought. Children of this age, four to eight years, will count in either way, and sometimes seem at a loss themselves to know where to begin. In one school room where this experiment was tried the teacher incautiously asked one child to count on his fingers, while all the other children in the room watched eagerly to see what he would do. He began with the little finger—and so did every child in the room after him. In another case the same error was made by the teacher, and the child first asked began with the thumb. Every other child in the room did the same, each following, consciously or unconsciously, the example of the leader. The results from these two schools were of course rejected from the totals which are given above; but they serve an excellent purpose in showing how slight is the preference which very young children have in this particular. So slight is it that no definite law can be postulated of this age; but the tendency seems to be to hold the palm of the hand downward, and then begin with the thumb. The writer once saw a boy about seven years old trying to multiply 3 by 6; and his method of procedure was as follows: holding his left hand with its palm down, he touched with the forefinger of his right hand the thumb, forefinger, and middle finger successively of his left hand. Then returning to his starting-point, he told off a second three in the same manner. This process he continued until he had obtained 6 threes, and then he announced his result correctly. If he had been a few years older, he might not have turned so readily to his thumb as a starting-point for any digital count. The indifference manifested by very young children gradually disappears, and at the age of twelve or thirteen the tendency is decidedly in the direction of beginning with the little finger. Fully three-fourths of all persons above that age will be found to count from the little finger toward the thumb, thus reversing the proportion that was found to obtain in the primary school rooms examined.

With respect to finger counting among civilized peoples, we fail, then, to find any universal law; the most that can be said is that more begin with the little finger than with the thumb. But when we proceed to the study of this slight but important particular among savages, we find them employing a certain order of succession with such substantial uniformity that the conclusion is inevitable that there must lie back of this some well-defined reason, or perhaps instinct, which guides them in their choice. This instinct is undoubtedly the outgrowth of the almost universal right-handedness of the human race. In finger counting, whether among children or adults, the beginning is made on the left hand, except in the case of left-handed individuals; and even then the start is almost as likely to be on the left hand as on the right. Savage tribes, as might be expected, begin with the left hand. Not only is this custom almost invariable, when tribes as a whole are considered, but the little finger is nearly always called into requisition first. To account for this uniformity, Lieutenant Gushing gives the following theory, [10] which is well considered, and is based on the results of careful study and observation among the Zuñi Indians of the Southwest: "Primitive man when abroad never lightly quit hold of his weapons. If he wanted to count, he did as the Zuñi afield does to-day; he tucked his instrument under his left arm, thus constraining the latter, but leaving the right hand free, that he might check off with it the fingers of the rigidly elevated left hand. From the nature of this position, however, the palm of the left hand was presented to the face of the counter, so that he had to begin his score on the little finger of it, and continue his counting from the right leftward. An inheritance of this may be detected to-day in the confirmed habit the Zuñi has of gesticulating from the right leftward, with the fingers of the right hand over those of the left, whether he be counting and summing up, or relating in any orderly manner." Here, then, is the reason for this otherwise unaccountable phenomenon. If savage man is universally right-handed, he will almost inevitably use the index finger of his right hand to mark the fingers counted, and he will begin his count just where it is most convenient. In his case it is with the little finger of the left hand. In the case of the child trying to multiply 3 by 6, it was with the thumb of the same hand. He had nothing to tuck under his arm; so, in raising his left hand to a position where both eye and counting finger could readily run over its fingers, he held the palm turned away from his face. The same choice of starting-point then followed as with the savage — the finger nearest his right hand; only in this case the finger was a thumb. The deaf mute is sometimes taught in this manner, which is for him an entirely natural manner. A left-handed child might be expected to count in a left-to-right manner, beginning, probably, with the thumb of his right hand.

To the law just given, that savages begin to count on the little finger of the left hand, there have been a few exceptions noted; and it has been observed that the method of progression on the second hand is by no means as invariable as on the first. The Otomacs [11] of South America began their count with the thumb, and to express the number 3 would use the thumb, forefinger, and middle finger. The Maipures, [12] oddly enough, seem to have begun, in some cases at least, with the forefinger; for they are reported as expressing 3 by means of the fore, middle, and ring fingers. The Andamans [13] begin with the little finger of either hand, tapping the nose with each finger in succession. If they have but one to express, they use the forefinger of either hand, pronouncing at the same time the proper word. The Bahnars, [14] one of the native tribes of the interior of Cochin China, exhibit no particular order in the sequence of fingers used, though they employ their digits freely to assist them in counting. Among certain of the negro tribes of South Africa [15] the little finger of the right hand is used for 1, and their count proceeds from right to left. With them, 6 is the thumb of the left hand, 7 the forefinger, and so on. They hold the palm downward instead of upward, and thus form a complete and striking exception to the law which has been found to obtain with such substantial uniformity in other parts of the uncivilized world. In Melanesia a few examples of preference for beginning with the thumb may also be noticed. In the Banks Islands the natives begin by turning down the thumb of the right hand, and then the fingers in succession to the little finger, which is 5. This is followed by the fingers of the left hand, both hands with closed fists being held up to show the completed 10. In Lepers' Island, they begin with the thumb, but, having reached 5 with the little finger, they do not pass to the other hand, but throw up the fingers they have turned down, beginning with the forefinger and keeping the thumb for 10. [16] In the use of the single hand this people is quite peculiar. The second 5 is almost invariably told off by savage tribes on the second hand, though in passing from the one to the other primitive man does not follow any invariable law. He marks 6 with either the thumb or the little finger. Probably the former is the more common practice, but the statement cannot be made with any degree of certainty. Among the Zulus the sequence is from thumb to thumb, as is the case among the other South African tribes just mentioned; while the Veis and numerous other African tribes pass from thumb to little finger. The Eskimo, and nearly all the American Indian tribes, use the correspondence between 6 and the thumb; but this habit is by no means universal. Respecting progression from right to left or left to right on the toes, there is no general law with which the author is familiar. Many tribes never use the toes in counting, but signify the close of the first 10 by clapping the hands together, by a wave of the right hand, or by designating some object; after which the fingers are again used as before.

One other detail in finger counting is worthy of a moment's notice. It seems to have been the opinion of earlier investigators that in his passage from one finger to the next, the savage would invariably bend down, or close, the last finger used; that is, that the count began with the fingers open and outspread. This opinion is, however, erroneous. Several of the Indian tribes of the West [17] begin with the hand clenched, and open the fingers one by one as they proceed. This method is much less common than the other, but that it exists is beyond question.

In the Muralug Island, in the western part of Torres Strait, a somewhat remarkable method of counting formerly existed, which grew out of, and is to be regarded as an extension of, the digital method. Beginning with the little finger of the left hand, the natives counted up to 5 in the usual manner, and then, instead of passing to the other hand, or repeating the count on the same fingers, they expressed the numbers from 6 to 10 by touching and naming successively the left wrist, left elbow, left shoulder, left breast, and sternum. Then the numbers from 11 to 19 were indicated by the use, in inverse order, of the corresponding portions of the right side, arm, and hand, the little finger of the right hand signifying 19. The words used were in each case the actual names of the parts touched; the same word, for example, standing for 6 and 14; but they were never used in the numerical sense unless accompanied by the proper gesture, and bear no resemblance to the common numerals, which are but few in number. This method of counting is rapidly dying out among the natives of the island, and is at the present time used only by old people. [18] Variations on this most unusual custom have been found to exist in others of the neighbouring islands, but none were exactly similar to it. One is also reminded by it of a custom [19] which has for centuries prevailed among bargainers in the East, of signifying numbers by touching the joints of each other's fingers under a cloth. Every joint has a special signification; and the entire system is undoubtedly a development from finger counting. The buyer or seller will by this method express 6 or 60 by stretching out the thumb and little finger and closing the rest of the fingers. The addition of the fourth finger to the two thus used signifies 7 or 70; and so on. "It is said that between two brokers settling a price by thus snipping with the fingers, cleverness in bargaining, offering a little more, hesitating, expressing an obstinate refusal to go further, etc., are as clearly indicated as though the bargaining were being carried on in words.

The place occupied, in the intellectual development of man, by finger counting and by the many other artificial methods of reckoning,—pebbles, shells, knots, the abacus, etc.,—seems to be this: The abstract processes of addition, subtraction, multiplication, division, and even counting itself, present to the mind a certain degree of difficulty. To assist in overcoming that difficulty, these artificial aids are called in; and, among savages of a low degree of development, like the Australians, they make counting possible. A little higher in the intellectual scale, among the American Indians, for example, they are employed merely as an artificial aid to what could be done by mental effort alone. Finally, among semi-civilized and civilized peoples, the same processes are retained, and form a part of the daily life of almost every person who has to do with counting, reckoning, or keeping tally in any manner whatever. They are no longer necessary, but they are so convenient and so useful that civilization can never dispense with them. The use of the abacus, in the form of the ordinary numeral frame, has increased greatly within the past few years; and the time may come when the abacus in its proper form will again find in civilized countries a use as common as that of five centuries ago.

In the elaborate calculating machines of the present, such as are used by life insurance actuaries and others having difficult computations to make, we have the extreme of development in the direction of artificial aid to reckoning. But instead of appearing merely as an extraneous aid to a defective intelligence, it now presents itself as a machine so complex that a high degree of intellectual power is required for the mere grasp of its construction and method of working.

Chapter II.

Number System Limits.

With respect to the limits to which the number systems of the various uncivilized races of the earth extend, recent anthropological research has developed many interesting facts. In the case of the Chiquitos and a few other native races of Bolivia we found no distinct number sense at all, as far as could be judged from the absence, in their language, of numerals in the proper sense of the word. How they indicated any number greater than *one* is a point still requiring investigation. In all other known instances we find actual number systems, or what may for the sake of uniformity be dignified by that name. In many cases, however, the numerals existing are so few, and the ability to count is so limited, that the term *number system* is really an entire misnomer.

Among the rudest tribes, those whose mode of living approaches most nearly to utter savagery, we find a certain uniformity of method. The entire number system may consist of but two words, *one* and *many*; or of three words, *one, two, many*. Or, the count may proceed to 3, 4, 5, 10, 20, or 100; passing always, or almost always, from the distinct numeral limit to the indefinite *many* or several, which serves for the expression of any number not readily grasped by the mind. As a matter of fact, most races count as high as 10; but to this statement the exceptions are so numerous that they deserve examination in some detail. In certain parts of the world, notably among the native races of South America, Australia, and many of the islands of Polynesia and Melanesia, a surprising paucity of numeral words has been observed. The Encabellada of the Rio Napo have but two distinct numerals; *tey*, 1, and *cayapa*, 2. [20] The Chaco languages [21] of the Guaycuru stock are also notably poor in this respect. In the Mbocobi dialect of this language the only native numerals are *yña tvak*, 1, and *yfioaca*, 2. The Puris [22] count *omi*, 1, *curiri*, 2, *prica*, many; and the Botocudos [23] *mokenam*, 1, *uruhu*, many. The Fuegans, [24] supposed to have been able at one time to count to 10, have but three numerals,—*kaoueli*, 1, *compaipi*, 2, *maten*, 3. The Campas of

Peru [25] possess only three separate words for the expression of number,—*patrio*, 1, *pitteni*, 2, *mahuani*, 3. Above 3 they proceed by combinations, as 1 and 3 for 4, 1 and 1 and 3 for 5. Counting above 10 is, however, entirely inconceivable to them, and any number beyond that limit they indicate by *tohaine*, many. The Conibos, [26] of the same region, had, before their contact with the Spanish, only *atchoupre*, 1, and *rrabui*, 2; though they made some slight progress above 2 by means of reduplication. The Orejones, one of the low, degraded tribes of the Upper Amazon, [27] have no names for number except *nayhay*, 1, *nenacome*, 2, *feninichacome*, 3, *ononoeomere*, 4. In the extensive vocabularies given by Von Martins, [28] many similar examples are found. For the Bororos he gives only *couai*, 1, *maeouai*, 2, *ouai*, 3. The last word, with the proper finger pantomime, serves also for any higher number which falls within the grasp of their comprehension. The Guachi manage to reach 5, but their numeration is of the rudest kind, as the following scale shows: *tamak*, 1, *eu-echo*, 2, *eu-echo-kailau*, 3, *eu-echo-way*, 4, *localau*, 5. The Carajas counted by a scale equally rude, and their conception of number seemed equally vague, until contact with the neighbouring tribes furnished them with the means of going beyond their original limit. Their scale shows clearly the uncertain, feeble number sense which is so marked in the interior of South America. It contains *wadewo*, 1, *wadebothoa*, 2, *wadeboaheodo*, 3, *wadebojeodo*, 4, *wadewajouclay*, 5, *wadewasori*, 6, or many.

Turning to the languages of the extinct, or fast vanishing, tribes of Australia, we find a still more noteworthy absence of numeral expressions. In the Gudang dialect [29] but two numerals are found— *pirman*, 1, and *ilabiu*, 2; in the Weedookarry, *ekkamurda*, 1, and *kootera*, 2; and in the Queanbeyan, *midjemban*, 1, and *bollan*, 2. In a score or more of instances the numerals stop at 3. The natives of Keppel Bay count *webben*, 1, *booli*, 2, *koorel*, 3; of the Boyne River, *karroon*, 1, *boodla*, 2, *numma*, 3; of the Flinders River, *kooroin*, 1, *kurto*, 2, *kurto kooroin*, 3; at the mouth of the Norman River, *lum*, 1, *buggar*, 2, *orinch*, 3; the Eaw tribe, *koothea*, 1, *woother*, 2, *marronoo*, 3; the Moree, *mal*, 1, *boolar*, 2, *kooliba*, 3; the Port Essington, [30] *erad*, 1, *nargarick*, 2, *nargarickelerad*, 3; the Darnly Islanders, [31] *netat*, 1, *naes*, 2, *naesa netat*, 3; and so on through a long list of tribes whose numeral scales are equally scanty. A still larger number of tribes show an ability to count one step further, to 4; but beyond this limit the majority of Australian and Tasmanian tribes do not go. It seems most remarkable that any human being should possess the ability to count to 4, and not to 5. The number of fingers on one hand furnishes so obvious a limit to any of these rudimentary systems, that positive evidence is needed before one can accept the statement. A careful examination of the numerals in upwards of a hundred Australian dialects leaves no doubt, however, that such is the fact. The Australians in almost all cases count by pairs; and so pronounced is this tendency that they pay but little attention to the fingers. Some tribes do not appear ever to count beyond 2 — a single pair. Many more go one step further; but if they do, they are as likely as not to designate their next numeral as two-one, or possibly, one-two. If this step is taken, we may or may not find one more added to it, thus completing the second pair. Still, the Australian's capacity for understanding anything which pertains to number is so painfully limited that even here there is sometimes an indefinite expression formed, as many, heap, or plenty, instead of any distinct numeral; and it is probably true that no Australian language contains a pure, simple numeral for 4. Curr, the best authority on this subject, believes that, where a distinct word for 4 is given, investigators have been deceived in every case. [32] If counting is carried beyond 4, it is always by means of reduplication. A few tribes gave expressions for 5, fewer still for 6, and a very small number appeared able to reach 7. Possibly the ability to count extended still further; but if so, it consisted undoubtedly in reckoning one pair after another, without any consciousness whatever of the sum total save as a larger number.

The numerals of a few additional tribes will show clearly that all distinct perception of number is lost as soon as these races attempt to count above 3, or at most, 4. The Yuckaburra [33] natives can go no further than *wigsin*, 1, *bullaroo*, 2, *goolbora*, 3. Above here all is referred to as *moorgha*, many. The Marachowies [34] have but three distinct numerals, — *cooma*, 1, *cootera*, 2, *murra*, 3. For 4 they say *minna*, many. At Streaky Bay we find a similar list, with the same words, *kooma* and *kootera*, for 1 and 2, but entirely different terms, *karboo* and *yalkata* for 3 and many. The same method obtains in the Minnal Yungar tribe, where the only numerals are *kain*, 1, *kujal*, 2, *moa*, 3, and *bulla*, plenty. In the Pinjarra dialect we find *doombart*, 1, *gugal*, 2, *murdine*, 3, *boola*, plenty; and in the dialect described as belonging to "Eyre's Sand Patch," three definite terms are given—*kean*, 1, *koojal*, 2, *yalgatta*, 3, while a fourth, *murna*, served to describe anything greater. In all these examples the fourth numeral is indefinite; and the same statement is true of many other Australian languages. But more commonly still we find 4, and perhaps 3 also, expressed by reduplication. In the Port Mackay dialect [35] the latter numeral is compound, the count being *warpur*, 1, *boolera*, 2, *boolera warpur*, 3. For 4 the term is not given. In the dialect which prevailed between the Albert and Tweed rivers [36] the scale appears as *yaburu*, 1, *boolaroo*, 2, *boolaroo yaburu*, 3, and *gurul* for 4 or anything beyond. The Wiraduroi [37] have *numbai*, 1, *bula*, 2, *bula numbai*, 3, *bungu*, 4, or many, and *bungu galan* or *bian galan*, 5, or very many. The Kamilaroi [38] scale is still more irregular, compounding above 4 with little apparent method. The numerals are *mal*, 1, *bular*, 2, *guliba*, 3, *bular bular*, 4, *bular guliba*, 5, *guliba guliba*, 6. The last two numerals show that 5 is to these natives simply 2-3, and 6 is 3-3. For additional examples of a similar nature the extended list of Australian scales given in Chapter V. may be consulted.

Taken as a whole, the Australian and Tasmanian tribes seem to have been distinctly inferior to those of South America in their ability to use and to comprehend numerals. In all but two or three cases the Tasmanians [39] were found to be unable to proceed beyond 2; and as the foregoing examples have indicated, their Australian neighbours were but little better off. In one or two instances we do find Australian numeral scales which reach 10, and perhaps we may safely say 20. One of these is given in full in a subsequent chapter, and its structure gives rise to the suspicion that it was originally as limited as those of kindred tribes, and that it underwent a considerable development after the natives had come in contact with the Europeans. There is good reason to believe that no Australian in his wild state could ever count intelligently to 7. [40]

In certain portions of Asia, Africa, Melanesia, Polynesia, and North America, are to be found races whose number systems are almost and sometimes quite as limited as are those of the South. American and Australian tribes already cited, but nowhere else do we find these so abundant as in the two continents just mentioned, where example after example might be cited of tribes whose ability to count is circumscribed within the narrowest limits. The Veddas [41] of Ceylon have but two numerals, *ekkameī*, 1, *dekkameï*, 2. Beyond this they count *otameekaï, otameekaï, otameekaï*, etc.; *i.e.* "and one more, and one more, and one more," and so on indefinitely. The Andamans, [42] inhabitants of a group of islands in the Bay of Bengal, are equally limited in their power of counting. They have *ubatulda*, 1, and *ikporda*, 2; but they can go no further, except in a manner similar to that of the Veddas. Above two they proceed wholly by means of the fingers, saying as they tap the nose with each successive finger, *anka*, "and this." Only the more intelligent of the Andamans can count at all, many of them seeming to be as nearly destitute of the number sense as it is possible for a human being to be. The Bushmen [43] of South Africa have but two numerals, the pronunciation of which can hardly be indicated without other resources than those of the English alphabet. Their word for 3 means, simply, many, as in the case of some of the Australian tribes. The Watchandies [44] have but two simple numerals, and their entire number system is *cooteon*, 1, *utaura*, 2, *utarra cooteoo*, 3, *atarra utarra*, 4. Beyond this they can only say, *booltha*, many, and *booltha bat*, very many. Although they have the expressions here given for 3 and 4, they are reluctant to use them, and only do so when absolutely required. The natives of Lower California [45] cannot count above 5. A few of the more intelligent among them understand the meaning of 2 fives, but this number seems entirely beyond the comprehension of the ordinary native. The Comanches, curiously enough, are so reluctant to employ their number words that they appear to prefer finger pantomime instead, thus giving rise to the impression which at one time became current, that they had no numerals at all for ordinary counting.

Aside from the specific examples already given, a considerable number of sweeping generalizations may be made, tending to show how rudimentary the number sense may be in aboriginal life. Scores of the native dialects of Australia and South America have been found containing number systems but little more extensive than those alluded to above. The negro tribes of Africa give the same testimony, as do many of the native races of Central America, Mexico, and the Pacific coast of the United States and Canada, the northern part of Siberia, Greenland, Labrador, and the arctic archipelago.

In speaking of the Eskimos of Point Barrow, Murdoch [46] says: "It was not easy to obtain any accurate information about the numeral system of these people, since in ordinary conversation they are not in the habit of specifying any numbers above five." Counting is often carried higher than this among certain of these northern tribes, but, save for occasional examples, it is limited at best. Dr. Franz Boas, who has travelled extensively among the Eskimos, and whose observations are always of the most accurate nature, once told the author that he never met an Eskimo who could count above 15. Their numerals actually do extend much higher; and a stray numeral of Danish origin is now and then met with, showing that the more intelligent among them are able to comprehend numbers of much greater magnitude than this. But as Dr. Boas was engaged in active work among them for three years, we may conclude that the Eskimo has an arithmetic but little more extended than that which sufficed for the Australians and the forest tribes of Brazil. Early Russian explorers among the northern tribes of Siberia noticed the same difficulty in ordinary, every-day reckoning among the natives. At first thought we might, then, state it as a general law that those races which are lowest in the scale of civilization, have the feeblest number sense also; or in other words, the least possible power of grasping the abstract idea of number.

But to this law there are many and important exceptions. The concurrent testimony of explorers seems to be that savage races possess, in the great majority of cases, the ability to count at least as high as 10. This limit is often extended to 20, and not infrequently to 100. Again, we find 1000 as the limit; or perhaps 10,000; and sometimes the savage carries his number system on into the hundreds of thousands or millions. Indeed, the high limit to which some savage races carry their numeration is far more worthy of remark than the entire absence of the number sense exhibited by others of apparently equal intelligence. If the life of any tribe is such as to induce trade and barter with their neighbours, a considerable quickness in reckoning will be developed among them. Otherwise this power will remain dormant because there is but little in the ordinary life of primitive man to call for its exercise.

In giving 1, 2, 3, 5, 10, or any other small number as a system limit, it must not be overlooked that this limit mentioned is in all cases the limit of the spoken numerals at the savage's command. The actual ability to count is almost always, and one is tempted to say always, somewhat greater than their vocabularies would indicate. The Bushman has no number word that will express for him anything higher than 2; but with the assistance of his fingers he gropes his way on as far as 10. The Veddas, the Andamans, the Guachi, the Botocudos, the Eskimos, and the thousand and one other tribes which furnish such scanty numeral systems, almost all proceed with more or less readiness as far as their fingers will carry them. As a matter of fact, this limit is frequently extended to 20; the toes, the fingers of a second man, or a recount of the savage's own fingers, serving as a tale for the second 10. Allusion is again made to this in a later chapter, where the subject of counting on the fingers and toes is examined more in detail.

In saying that a savage can count to 10, to 20, or to 100, but little idea is given of his real mental conception of any except the smallest numbers. Want of familiarity with the use of numbers, and lack of convenient means of comparison, must result in extreme indefiniteness of mental conception and almost entire absence of exactness. The experience of Captain Parry, [47] who found that the Eskimos made mistakes before they reached 7, and of Humboldt, [48] who says that a Chayma might be made to say that his age was either 18 or 60, has been duplicated by all investigators who have had actual experience among savage races. Nor, on the other hand, is the development of a numeral system an infallible index of mental power, or of any real approach toward civilization. A continued use of the trading and bargaining faculties must and does result in a familiarity with numbers sufficient to enable savages to perform unexpected feats in reckoning. Among some of the West African tribes this has actually been found to be the case; and among the Yorubas of Abeokuta [49] the extraordinary saying, "You may seem very clever, but you can't tell nine times nine," shows how surprisingly this faculty has been developed, considering the general condition of savagery in which the tribe lived. There can be no doubt that, in general, the growth of the number sense keeps pace with the growth of the intelligence in other respects. But when it is remembered that the Tonga Islanders have numerals up to 100,000, and the Tembus, the Fingoes, the Pondos, and a dozen other South African tribes go as high as 1,000,000; and that Leigh Hunt never could learn the multiplication table, one must confess that this law occasionally presents to our consideration remarkable exceptions.

While considering the extent of the savage's arithmetical knowledge, of his ability to count and to grasp the meaning of number, it may not be amiss to ask ourselves the question, what is the extent of the development of our own number sense? To what limit can we absorb the idea of number, with a complete appreciation of the idea of the number of units involved in any written or spoken quantity? Our perfect system of numeration enables us to express without difficulty any desired number, no matter how great or how small it be. But how much of actually clear comprehension does the number thus expressed convey to the mind? We say that one place is 100 miles from another; that A paid B 1000 dollars for a certain piece of property; that a given city contains 10,000 inhabitants; that 100,000 bushels of wheat were shipped from Duluth or Odessa on such a day; that 1,000,000 feet of lumber were destroyed by the fire of yesterday,—and as we pass from the smallest to the largest of the numbers thus instanced, and from the largest on to those still larger, we repeat the question just asked; and we repeat it with a new sense of our own mental limitation. The number 100 unquestionably stands for a distinct conception. Perhaps the same may be said for 1000, though this could not be postulated with equal certainty. But what of 10,000? If that number of persons were gathered together into a single hall or amphitheatre, could an estimate be made by the average onlooker which would approximate with any degree of accuracy the size of the assembly? Or if an observer were stationed at a certain point, and 10,000 persons were to pass him in single file without his counting them as they passed, what sort of an estimate would he make of their number? The truth seems to be that our mental conception of number is much more limited than is commonly thought, and that we unconsciously adopt some new unit as a standard of comparison when we wish to render intelligible to our minds any number of considerable magnitude. For example, we say that A has a fortune of $1,000,000. The impression is at once conveyed of a considerable degree of wealth, but it is rather from the fact that that fortune represents an annual income of $40,000 than, from the actual magnitude of the fortune itself. The number 1,000,000 is, in itself, so greatly in excess of anything that enters into our daily experience that we have but a vague conception of it, except as something very great. We are not, after all, so very much better off than the child who, with his arms about his mother's neck, informs her with perfect gravity and sincerity that he "loves her a million bushels." His idea is merely of some very great amount, and our own is often but little clearer when we use the expressions which are so easily represented by a few digits. Among the uneducated portions of civilized communities the limit of clear comprehension of number is not only relatively, but absolutely, very low. Travellers in Russia have informed the writer that the peasants of that country have no distinct idea of a number consisting of but a few hundred even. There is no reason to doubt this testimony. The entire life of a peasant might be passed without his ever having occasion to use a number as great as 500, and as a result he might have respecting that number an idea less distinct than a trained mathematician would have of the distance from the earth to the sun. De Quincey [50] incidentally mentions this characteristic in narrating a conversation which occurred while he was at Carnarvon, a little town in Wales. "It was on this occasion," he says, "that I learned how vague are the ideas of number in unpractised minds. 'What number of people do you think,' I said to an elderly person, 'will be assembled this day at Carnarvon?' 'What number?' rejoined the person addressed; 'what number? Well, really, now, I should reckon—perhaps a matter of four million.' Four millions of *extra* people in little Carnarvon, that could barely find accommodation (I should calculate) for an extra four hundred!" So the Eskimo and the South American Indian are, after all, not so very far behind the "elderly person" of Carnarvon, in the distinct perception of a number which familiarity renders to us absurdly small.

Chapter III.

The Origin of Number Words.

In the comparison of languages and the search for primitive root forms, no class of expressions has been subjected to closer scrutiny than the little cluster of words, found in each language, which constitutes a part of the daily vocabulary of almost every human being—the words with which we begin our counting. It is assumed, and with good reason, that these are among the earlier words to appear in any language; and in the mutations of human speech, they are found to suffer less than almost any other portion of a language. Kinship between tongues remote from each other has in many instances been detected by the similarity found to exist among the every-day words of each; and among these words one may look with a good degree of certainty for the 1, 2, 3, etc., of the number scale. So fruitful has been this line of research, that the attempt has been made, even, to establish a common origin for all the races of mankind by means of a comparison of numeral words. [51] But in this instance, as in so many others that will readily occur to the mind, the result has been that the theory has finally taken possession of the author and reduced him to complete subjugation, instead of remaining his servant and submitting to the legitimate results of patient and careful investigation. Linguistic research is so full of snares and pitfalls that the student must needs employ the greatest degree of discrimination before asserting kinship of race because of resemblances in vocabulary; or even relationship between words in the same language because of some chance likeness of form that may exist between them. Probably no one would argue that the English and the Babusessé of Central Africa were of the same primitive stock simply because in the language of the latter *five atano* means 5, and *ten kumi* means 10. [52] But, on the other hand, many will argue that, because the German *zehn* means 10, and *zehen* means toes, the ancestors of the Germans counted on their toes; and that with them, 10 was the complete count of the toes. It may be so. We certainly have no evidence with which to disprove this; but, before accepting it as a fact, or even as a reasonable hypothesis, we may be pardoned for demanding some evidence aside from the mere resemblance in the form of the words. If, in the study of numeral words, form is to constitute our chief guide, we must expect now and then to be confronted with facts which are not easily reconciled with any pet theory.

The scope of the present work will admit of no more than a hasty examination of numeral forms, in which only actual and well ascertained meanings will be considered. But here we are at the outset confronted with a class of words whose original meanings appear to be entirely lost. They are what may be termed the numerals proper—the native, uncompounded words used to signify number. Such words are the one, two, three, etc., of English; the eins, zwei, drei, etc., of German; words which must at some time, in some prehistoric language, have had definite meanings entirely apart from those which they now convey to our minds. In savage languages it is sometimes possible to detect these meanings, and thus to obtain possession of the clue that leads to the development, in the barbarian's rude mind, of a count scale—a number system. But in languages like those of modern Europe, the pedigree claimed by numerals is so long that, in the successive changes through which they have passed, all trace of their origin seems to have been lost.

The actual number of such words is, however, surprisingly small in any language. In English we count by simple words only to 10. From this point onward all our numerals except "hundred" and "thousand" are compounds and combinations of the names of smaller numbers. The words we employ to designate the higher orders of units, as million, billion, trillion, etc., are appropriated bodily from the Italian; and the native words *pair, tale, brace, dozen, gross,* and *score,* can hardly be classed as numerals in the strict sense of the word. German possesses exactly the same number of native words in its numeral scale as English; and the same may be said of the Teutonic languages generally, as well as of the Celtic, the Latin, the Slavonic, and the Basque. This is, in fact, the universal method observed in the formation of any numeral scale, though the actual number of simple words may vary. The Chiquito language has but one numeral of any kind whatever; English contains twelve simple terms; Sanskrit has twenty-seven, while Japanese possesses twenty-four, and the Chinese a number almost equally great. Very many languages, as might be expected, contain special numeral expressions, such as the German *dutzend* and the French *dizaine*; but these, like the English *dozen* and *score*, are not to be regarded as numerals proper.

The formation of numeral words shows at a glance the general method in which any number scale has been built up. The primitive savage counts on his fingers until he has reached the end of one, or more probably of both, hands. Then, if he wishes to proceed farther, some mark is made, a pebble is laid aside, a knot tied, or some similar device employed to signify that all the counters at his disposal have been used. Then the count begins anew, and to avoid multiplication of words, as well as to assist the memory, the terms already used are again resorted to; and the name by which the first halting-place was designated is repeated with each new numeral. Hence the thirteen, fourteen, fifteen, etc., which are contractions of the fuller expressions three-and-ten, four-and-ten, five-and-ten, etc. The specific method of combination may not always be the same, as witness the *eighteen*, or eight-ten, in English, and *dix-huit*, or ten-eight, in French; *forty-five*, or four-tens-five, in English, and *fünf und vierzig*, or five and four tens in German. But the general method is the same the world over, presenting us with nothing but local variations, which are, relatively speaking, entirely unimportant. With this fact in mind, we can cease to wonder at the small number of simple numerals in any language. It might, indeed, be queried, why do any languages, English and German, for example, have unusual compounds for 11 and 12? It would seem as though the regular method of compounding should begin with 10 and 1, instead of 10 and 3, in any language using a system with 10 as a base. An examination of several hundred numeral scales shows that the Teutonic languages are somewhat exceptional in this respect. The words *eleven* and *twelve* are undoubtedly combinations, but not in the same direct sense as *thirteen, twenty-five*, etc. The same may be said of the French *onze, douze, treize, quatorze, quinze*, and *seize*, which are obvious compounds, but not formed in the same manner as the numerals above that point. Almost all civilized languages, however, except the Teutonic, and practically all uncivilized languages, begin their direct numeral combinations as soon as they have passed their number base, whatever that may be. To give an illustration, selected quite at random from among the barbarous tribes of Africa, the Ki-Swahili numeral scale runs as follows: [53]

1.	moyyi,
2.	mbiri,
3.	tato,
4.	ena,
5.	tano,
6.	seta,
7.	saba,
8.	nani,
9.	kenda,
10.	kumi,
11.	kumi na moyyi,
12.	kumi na mbiri,
13.	kumi na tato,
	etc.

The words for 11, 12, and 13, are seen at a glance to signify ten-and-one, ten-and-two, ten-and-three, and the count proceeds, as might be inferred, in a similar manner as far as the number system extends. Our English combinations are a little closer than these, and the combinations found in certain other languages are, in turn, closer than those of the English; as witness the *once*, 11, *doce*, 12, *trece*, 13, etc., of Spanish. But the process is essentially the same, and the law may be accepted as practically invariable, that all numerals greater than the base of a system are expressed by compound words, except such as are necessary to establish some new order of unit, as hundred or thousand.

In the scale just given, it will be noticed that the larger number precedes the smaller, giving 10 + 1, 10 + 2, etc., instead of 1 + 10, 2 + 10, etc. This seems entirely natural, and hardly calls for any comment whatever. But we have only to consider the formation of our English "teens" to see that our own method is, at its inception, just the reverse of this. Thirteen, 14, and the remaining numerals up to 19 are formed by prefixing the smaller number to the base; and it is only when we pass 20 that we return to the more direct and obvious method of giving precedence to the larger. In German and other Teutonic languages the inverse method is continued still further. Here 25 is *fünf und zwanzig*, 5 and 20; 92 is *zwei und neunzig*, 2 and 90, and so on to 99. Above 100 the order is made direct, as in English. Of course, this mode of formation between 20 and 100 is permissible in English, where "five and twenty" is just as correct a form as twenty-five. But it is archaic, and would soon pass out of the language altogether, were it not for the influence of some of the older writings which have had a strong influence in preserving for us many of older and more essentially Saxon forms of expression.

Both the methods described above are found in all parts of the world, but what I have called the direct is far more common than the other. In general, where the smaller number precedes the larger it signifies multiplication instead of addition. Thus, when we say "thirty," *i.e.* three-ten, we mean 3×10; just as "three hundred" means 3×100. When the larger precedes the smaller, we must usually understand addition. But to both these rules there are very many exceptions. Among higher numbers the inverse order is very rarely used; though even here an occasional exception is found. The Taensa Indians, for example, place the smaller numbers before the larger, no matter how far their scale may extend. To say 1881 they make a complete inversion of our own order, beginning with 1 and ending with 1000. Their full numeral for this is *yeha av wabki mar-u-wab mar-u-haki*, which means, literally, $1 + 80 + 100 \times 8 + 100 \times 10$. [54] Such exceptions are, however, quite rare.

One other method of combination, that of subtraction, remains to be considered. Every student of Latin will recall at once the *duodeviginti*, 2 from 20, and *undeviginti*, 1 from 20, which in that language are the regular forms of expression for 18 and 19. At first they seem decidedly odd; but familiarity soon accustoms one to them, and they cease entirely to attract any special attention. This principle of subtraction, which, in the formation of numeral words, is quite foreign to the genius of English, is still of such common occurrence in other languages that the Latin examples just given cease to be solitary instances.

The origin of numerals of this class is to be found in the idea of reference, not necessarily to the last, but to the nearest, halting-point in the scale. Many tribes seem to regard 9 as "almost 10," and to give it a name which conveys this thought. In the Mississaga, one of the numerous Algonquin languages, we have, for example, the word *cangaswi*, "incomplete 10," for 9. [55] In the Kwakiutl of British Columbia, 8 as well as 9 is formed in this way; these two numbers being *matlguanatl*, 10 − 2, and *nanema*, 10 − 1, respectively. [56] In many of the languages of British Columbia we find a similar formation for 8 and 9, or for 9 alone. The same formation occurs in Malay, resulting in the numerals *delapan*, 10 − 2, and *sambilan* 10 − 1. [57] In Green Island, one of the New Ireland group, these become simply *andra-lua*, "less 2," and *andra-si*, "less 1." [58] In the Admiralty Islands this formation is carried back one step further, and not only gives us *shua-luea*, "less 2," and *shu-ri*, "less 1," but also makes 7 appear as *sua-tolu*, "less 3." [59] Surprising as this numeral is, it is more than matched by the Ainu scale, which carries subtraction back still another step, and calls 6, 10 − 4. The four numerals from 6 to 9 in this scale are respectively, *iwa*, 10 − 4, *arawa*, 10 − 3, *tupe-san*, 10 − 2, and *sinepe-san*, 10 − 1. [60] Numerous examples of this kind of formation will be found in later chapters of this work; but they will usually be found to occur in one or both of the numerals, 8 and 9. Occasionally they appear among the higher numbers; as in the Maya languages, where, for example, 99 years is "one single year lacking from five score years," [61] and in the Arikara dialects, where 98 and 99 are "5 men minus" and "5 men 1 not." [62] The Welsh, Danish, and other languages less easily accessible than these to the general student, also furnish interesting examples of a similar character.

More rarely yet are instances met with of languages which make use of subtraction almost as freely as addition, in the composition of numerals. Within the past few years such an instance has been noticed in the case of the Bellacoola language of British Columbia. In their numeral scale 15, "one foot," is followed by 16, "one man less 4"; 17, "one man less 3"; 18, "one man less 2"; 19, "one man less 1"; and 20, one man. Twenty-five is "one man and one hand"; 26, "one man and two hands less 4"; 36, "two men less 4"; and so on. This method of formation prevails throughout the entire numeral scale. [63]

One of the best known and most interesting examples of subtraction as a well-defined principle of formation is found in the Maya scale. Up to 40 no special peculiarity appears; but as the count progresses beyond that point we find a succession of numerals which one is almost tempted to call 60 − 19, 60 − 18, 60 − 17, etc. Literally translated the meanings seem to be 1 to 60, 2 to 60, 3 to 60, etc. The point of reference is 60, and the thought underlying the words may probably be expressed by the paraphrases, "1 on the third score, 2 on the third score, 3 on the third score," etc. Similarly, 61 is 1 on the fourth score, 81 is one on the fifth score, 381 is 1 on the nineteenth score, and so on to 400. At 441 the same formation reappears; and it continues to characterize the system in a regular and consistent manner, no matter how far it is extended. [64]

The Yoruba language of Africa is another example of most lavish use of subtraction; but it here results in a system much less consistent and natural than that just considered. Here we find not only 5, 10, and 20 subtracted from the next higher unit, but also 40, and even 100. For example, 360 is 400 − 40; 460 is 500 − 40; 500 is 600 − 100; 1300 is 1400 − 100, etc. One of the Yoruba units is 200; and all the odd hundreds up to 2000, the next higher unit, are formed by subtracting 100 from the next higher multiple of 200. The system is quite complex, and very artificial; and seems to have been developed by intercourse with traders. [65]

It has already been stated that the primitive meanings of our own simple numerals have been lost. This is also true of the languages of nearly all other civilized peoples, and of numerous savage races as well. We are at liberty to suppose, and we do suppose, that in very many cases these words once expressed meanings closely connected with the names of the fingers, or with the fingers themselves, or both. Now and then a case is met with in which the numeral word frankly avows its meaning — as in the Botocudo language, where 1 is expressed by *podzik*, finger, and 2 by *kripo*, double finger; [66] and in the Eskimo dialect of Hudson's Bay, where *eerkitkoka* means both 10 and little finger. [67] Such cases are, however, somewhat exceptional.

In a few noteworthy instances, the words composing the numeral scale of a language have been carefully investigated and their original meanings accurately determined. The simple structure of many of the rude languages of the world should render this possible in a multitude of cases; but investigators are too often content with the mere numerals themselves, and make no inquiry respecting their meanings. But the following exposition of the Zuñi scale, given by Lieutenant Gushing [68] leaves nothing to be desired:

1. töpinte = taken to start with.
2. kwilli = put down together with.
3. ha'ï = the equally dividing finger.
4. awite = all the fingers all but done with.
5. öpte = the notched off.

This finishes the list of original simple numerals, the Zuñi stopping, or "notching off," when he finishes the fingers of one hand. Compounding now begins.

6. topalïk'ya = another brought to add to the done with.
7. kwillilïk'ya = two brought to and held up with the rest.
8. hailïk'ye = three brought to and held up with the rest.
9. tenalïk'ya = all but all are held up with the rest.
10. ästem'thila = all the fingers.
11. ästem'thla topayä'th-l'tona = all the fingers and another over above held.

The process of formation indicated in 11 is used in the succeeding numerals up to 19.

20. kwillik'yënästem'thlan = two times all the fingers.
100. ässiästem'thlak'ya = the fingers all the fingers.
1000. ässiästem'thlanak'yënästem'thla = the fingers all the fingers times all the fingers.

The only numerals calling for any special note are those for 11 and 9. For 9 we should naturally expect a word corresponding in structure and meaning to the words for 7 and 8. But instead of the "four brought to and held up with the rest," for which we naturally look, the Zuñi, to show that he has used all of his fingers but one, says "all but all are held up with the rest." To express 11 he cannot use a similar form of composition, since he has already used it in constructing his word for 6, so he says "all the fingers and another over above held."

The one remarkable point to be noted about the Zuñi scale is, after all, the formation of the words for 1 and 2. While the savage almost always counts on his fingers, it does not seem at all certain that these words would necessarily be of finger formation. The savage can always distinguish between one object and two objects, and it is hardly reasonable to believe that any external aid is needed to arrive at a distinct perception of this difference. The numerals for 1 and 2 would be the earliest to be formed in any language, and in most, if not all, cases they would be formed long before the need would be felt for terms to describe any higher number. If this theory be correct, we should expect to find finger names for numerals beginning not lower than 3, and oftener with 5 than with any other number. The highest authority has ventured the assertion that all numeral words have their origin in the names of the fingers; [69] substantially the same conclusion was reached by Professor Pott, of Halle, whose work on numeral nomenclature led him deeply into the study of the origin of these words. But we have abundant evidence at hand to show that, universal as finger counting has been, finger origin for numeral words has by no means been universal. That it is more frequently met with than any other origin is unquestionably true; but in many instances, which will be more fully considered in the following chapter, we find strictly non-digital derivations, especially in the case of the lowest members of the scale. But in nearly all languages the origin of the words for 1, 2, 3, and 4 are so entirely unknown that speculation respecting them is almost useless.

An excellent illustration of the ordinary method of formation which obtains among number scales is furnished by the Eskimos of Point Barrow, [70] who have pure numeral words up to 5, and then begin a systematic course of word formation from the names of their fingers. If the names of the first five numerals are of finger origin, they have so completely lost their original form, or else the names of the fingers themselves have so changed, that no resemblance is now to be detected between them. This scale is so interesting that it is given with considerable fulness, as follows:

1. atauzik.

2. madro.

3. pinasun.

4. sisaman.

5. tudlemut.

6. atautyimin akbinigin [tudlimu(t)]   = 5 and 1 on the next.

7. madronin akbinigin   = twice on the next.

8. pinasunin akbinigin   = three times on the next.

9. kodlinotaila   = that which has not its 10.

10. kodlin   = the upper part — *i.e.* the fingers.

14. akimiaxotaityuna   = I have not 15.

15. akimia. [This seems to be a real numeral word.]

20. inyuina   = a man come to an end.

25. inyuina tudlimunin akbinidigin   = a man come to an end and 5 on the next.

30. inyuina kodlinin akbinidigin   = a man come to an end and 10 on the next.

35. inyuina akimiamin aipalin   = a man come to an end accompanied by 1 fifteen times.

40. madro inyuina   = 2 men come to an end.

In this scale we find the finger origin appearing so clearly and so repeatedly that one feels some degree of surprise at finding 5 expressed by a pure numeral instead of by some word meaning *hand* or *fingers of one hand*. In this respect the Eskimo dialects are somewhat exceptional among scales built up of digital words. The system of the Greenland Eskimos, though differing slightly from that of their Point Barrow cousins, shows the same peculiarity. The first ten numerals of this scale are: [71]

1. atausek.

2.   mardluk.

3.   pingasut.

4.   sisamat.

5.   tatdlimat.

6.   arfinek-atausek   = to the other hand 1.

7.   arfinek-mardluk   = to the other hand 2.

8.   arfinek-pingasut   = to the other hand 3.

9.   arfinek-sisamat   = to the other hand 4.

10.   kulit.

The same process is now repeated, only the feet instead of the hands are used; and the completion of the second 10 is marked by the word *innuk*, man. It may be that the Eskimo word for 5 is, originally, a digital word, but if so, the fact has not yet been detected. From the analogy furnished by other languages we are justified in suspecting that this may be the case; for whenever a number system contains digital words, we expect them to begin with *five*, as, for example, in the Arawak scale, [72] which runs:

1.   abba.

2.   biama.

3.   kabbuhin.

4.   bibiti.

5.   abbatekkábe   = 1 hand.

6.   abbatiman   = 1 of the other.

7.   biamattiman   = 2 of the other.

8.   kabbuhintiman   = 3 of the other.

9.   bibitiman   = 4 of the other.

10.   biamantekábbe   = 2 hands.

11.   abba kutihibena   = 1 from the feet.

20.   abba lukku   = hands feet.

The four sets of numerals just given may be regarded as typifying one of the most common forms of primitive counting; and the words they contain serve as illustrations of the means which go to make up the number scales of savage races. Frequently the finger and toe origin of numerals is perfectly apparent, as in the Arawak system just given, which exhibits the simplest and clearest possible method of formation. Another even more interesting system is that of the Montagnais of northern Canada. [73] Here, as in the Zuñi scale, the words are digital from the outset.

1.	inl'are	= the end is bent.
2.	nak'e	= another is bent.
3.	t'are	= the middle is bent.
4.	dinri	= there are no more except this.
5.	se-sunla-re	= the row on the hand.
6.	elkke-t'are	= 3 from each side.
7. {	t'a-ye-oyertan	= there are still 3 of them.
	inl'as dinri	= on one side there are 4 of them.
8.	elkke-dinri	= 4 on each side.
9.	inl'a-ye-oyert'an	= there is still 1 more.
10.	onernan	= finished on each side.
11.	onernan inl'are ttcharidhel	= 1 complete and 1.
12.	onernan nak'e ttcharidhel	= 1 complete and 2, etc.

The formation of 6, 7, and 8 of this scale is somewhat different from that ordinarily found. To express 6, the Montagnais separates the thumb and forefinger from the three remaining fingers of the left hand, and bringing the thumb of the right hand close to them, says: "3 from each side." For 7 he either subtracts from 10, saying: "there are still 3 of them," or he brings the thumb and forefinger of the right hand up to the thumb of the left, and says: "on one side there are 4 of them." He calls 8 by the same name as many of the other Canadian tribes, that is, two 4's; and to show the proper number of fingers, he closes the thumb and little finger of the right hand, and then puts the three remaining fingers beside the thumb of the left hand. This method is, in some of these particulars, different from any other I have ever examined.

It often happens that the composition of numeral words is less easily understood, and the original meanings more difficult to recover, than in the examples already given. But in searching for number systems which show in the formation of their words the influence of finger counting, it is not unusual to find those in which the derivation from native words signifying *finger, hand, toe, foot,* and *man,* is just as frankly obvious as in the case of the Zuñi, the Arawak, the Eskimo, or the Montagnais scale. Among the Tamanacs, [74] one of the numerous Indian tribes of the Orinoco, the numerals are as strictly digital as in any of the systems already examined. The general structure of the Tamanac scale is shown by the following numerals:

5.	amgnaitone	= 1 hand complete.
6.	itacono amgna pona tevinitpe	= 1 on the other hand.
10.	amgna aceponare	= all of the 2 hands.
11.	puitta pona tevinitpe	= 1 on the foot.
16.	itacono puitta pona tevinitpe	= 1 on the other foot.
20.	tevin itoto	= 1 man.
21.	itacono itoto jamgnar bona tevinitpe	= 1 on the hands of another man.

In the Guarani [75] language of Paraguay the same method is found, with a different form of expression for 20. Here the numerals in question are

5.	asepopetei	= one hand.
10.	asepomokoi	= two hands.
20.	asepo asepi abe	= hands and feet.

Another slight variation is furnished by the Kiriri language, [76] which is also one of the numerous South American Indian forms of speech, where we find the words to be

5. mi biche misa	= one hand.
10. mikriba misa sai	= both hands.
20. mikriba misa idecho ibi sai	= both hands together with the feet.

Illustrations of this kind might be multiplied almost indefinitely; and it is well to note that they may be drawn from all parts of the world. South America is peculiarly rich in native numeral words of this kind; and, as the examples above cited show, it is the field to which one instinctively turns when this subject is under discussion. The Zamuco numerals are, among others, exceedingly interesting, giving us still a new variation in method. They are [77]

1. tsomara.

2. gar.

3. gadiok.

4. gahagani.

5. tsuena yimana-ite   = ended 1 hand.

6. tsomara-hi   = 1 on the other.

7. gari-hi   = 2 on the other.

8. gadiog-ihi   = 3 on the other.

9. gahagani-hi   = 4 on the other.

10. tsuena yimana-die   = ended both hands.

11. tsomara yiri-tie   = 1 on the foot.

12. gar yiritie   = 2 on the foot.

20. tsuena yiri-die   = ended both feet.

As is here indicated, the form of progression from 5 to 10, which we should expect to be "hand-1," or "hand-and-1," or some kindred expression, signifying that one hand had been completed, is simply "1 on the other." Again, the expressions for 11, 12, etc., are merely "1 on the foot," "2 on the foot," etc., while 20 is "both feet ended."

An equally interesting scale is furnished by the language of the Maipures [78] of the Orinoco, who count

1. papita.

2. avanume.

3. apekiva.

4. apekipaki.

5. papitaerri capiti   = 1 only hand.

6. papita yana pauria capiti purena = 1 of the other hand we take.

10. apanumerri capiti   = 2 hands.

11. papita yana kiti purena   = 1 of the toes we take.

20. papita camonee	= 1 man.
40. avanume camonee	= 2 men.
60. apekiva camonee	= 3 men, etc.

In all the examples thus far given, 20 is expressed either by the equivalent of "man" or by some formula introducing the word "feet." Both these modes of expressing what our own ancestors termed a "score," are so common that one hesitates to say which is of the more frequent use. The following scale, from one of the Betoya dialects [79] of South America, is quite remarkable among digital scales, making no use of either "man" or "foot," but reckoning solely by fives, or hands, as the numerals indicate.

1.	tey.	
2.	cayapa.	
3.	toazumba.	
4.	cajezea	= 2 with plural termination.
5.	teente	= hand.
6.	teyentetey	= hand + 1.
7.	teyente cayapa	= hand + 2.
8.	teyente toazumba	= hand + 3.
9.	teyente caesea	= hand + 4.
10.	caya ente, or caya huena	= 2 hands.
11.	caya ente-tey	= 2 hands + 1.
15.	toazumba-ente	= 3 hands.
16.	toazumba-ente-tey	= 3 hands + 1.
20.	caesea ente	= 4 hands.

In the last chapter mention was made of the scanty numeral systems of the Australian tribes, but a single scale was alluded to as reaching the comparatively high limit of 20. This system is that belonging to the Pikumbuls, [80] and the count runs thus:

1.	mal.
2.	bular.
3.	guliba.

4.	bularbular	= 2-2.
5.	mulanbu.	
6.	malmulanbu mummi	= 1 and 5 added on.
7.	bularmulanbu mummi	= 2 and 5 added on.
8.	gulibamulanbu mummi	= 3 and 5 added on.
9.	bularbularmulanbu mummi	= 4 and 5 added on.
10.	bularin murra	= belonging to the 2 hands.
11.	maldinna mummi	= 1 of the toes added on (to the 10 fingers).
12.	bular dinna mummi	= 2 of the toes added on.
13.	guliba dinna mummi	= 3 of the toes added on.
14.	bular bular dinna mummi	= 4 of the toes added on.
15.	mulanba dinna	= 5 of the toes added on.
16.	mal dinna mulanbu	= 1 and 5 toes.
17.	bular dinna mulanbu	= 2 and 5 toes.
18.	guliba dinna mulanbu	= 3 and 5 toes.
19.	bular bular dinna mulanbu	= 4 and 5 toes.
20.	bularin dinna	= belonging to the 2 feet.

As has already been stated, there is good ground for believing that this system was originally as limited as those obtained from other Australian tribes, and that its extension from 4, or perhaps from 5 onward, is of comparatively recent date.

A somewhat peculiar numeral nomenclature is found in the language of the Klamath Indians of Oregon. The first ten words in the Klamath scale are: [81]

1.	nash, or nas.	
2.	lap	= hand.
3.	ndan.	
4.	vunep	= hand up.
5.	tunep	= hand away.

6.	nadshkshapta	= 1 I have bent over.
7.	lapkshapta	= 2 I have bent over.
8.	ndankshapta	= 3 I have bent over.
9.	nadshskeksh	= 1 left over.
10.	taunep	= hand hand?

In describing this system Mr. Gatschet says: "If the origin of the Klamath numerals is thus correctly traced, their inventors must have counted only the four long fingers without the thumb, and 5 was counted while saying *hand away! hand off!* The 'four,' or *hand high! hand up!* intimates that the hand was held up high after counting its four digits; and some term expressing this gesture was, in the case of *nine*, substituted by 'one left over' ... which means to say, 'only one is left until all the fingers are counted.'" It will be observed that the Klamath introduces not only the ordinary finger manipulation, but a gesture of the entire hand as well. It is a common thing to find something of the kind to indicate the completion of 5 or 10, and in one or two instances it has already been alluded to. Sometimes one or both of the closed fists are held up; sometimes the open hand, with all the fingers extended, is used; and sometimes an entirely independent gesture is introduced. These are, in general, of no special importance; but one custom in vogue among some of the prairie tribes of Indians, to which my attention was called by Dr. J. Owen Dorsey, [82] should be mentioned. It is a gesture which signifies multiplication, and is performed by throwing the hand to the left. Thus, after counting 5, a wave of the hand to the left means 50. As multiplication is rather unusual among savage tribes, this is noteworthy, and would seem to indicate on the part of the Indian a higher degree of intelligence than is ordinarily possessed by uncivilized races.

In the numeral scale as we possess it in English, we find it necessary to retain the name of the last unit of each kind used, in order to describe definitely any numeral employed. Thus, fifteen, one hundred forty-two, six thousand seven hundred twenty-seven, give in full detail the numbers they are intended to describe. In primitive scales this is not always considered necessary; thus, the Zamucos express their teens without using their word for 10 at all. They say simply, 1 on the foot, 2 on the foot, etc. Corresponding abbreviations are often met; so often, indeed, that no further mention of them is needed. They mark one extreme, the extreme of brevity, found in the savage method of building up hand, foot, and finger names for numerals; while the Zuñi scale marks the extreme of prolixity in the formation of such words. A somewhat ruder composition than any yet noticed is shown in the numerals of the Vilelo scale, [83] which are:

1. agit, or yaagit.

2. uke.

3. nipetuei.

4. yepkatalet.

5. isig-nisle-yaagit   = hand fingers 1.

6. isig-teet-yaagit   = hand with 1.

7. isig-teet-uke   = hand with 2.

8. isig-teet-nipetuei   = hand with 3.

9. isig-teet-yepkatalet   = hand with 4.

10. isig-uke-nisle   = second hand fingers (lit. hand-two-fingers).

11. isig-uke-nisle-teet-yaagit   = second hand fingers with 1.

20. isig-ape-nisle-lauel   = hand foot fingers all.

In the examples thus far given, it will be noticed that the actual names of individual fingers do not appear. In general, such words as thumb, forefinger, little finger, are not found, but rather the hand-1, 1 on the next, or 1 over and above, which we have already seen, are the type forms for which we are to look. Individual finger names do occur, however, as in the scale of the Hudson's Bay Eskimos, [84] where the three following words are used both as numerals and as finger names:

8. kittukleemoot   = middle finger.

9. mikkeelukkamoot   = fourth finger.

10. eerkitkoka   = little finger.

Words of similar origin are found in the original Jiviro scale, [85] where the native numerals are:

1. ala.
2. catu.
3. cala.
4. encatu.
5. alacötegladu   = 1 hand.
6. intimutu   = thumb (of second hand).
7. tannituna   = index finger.
8. tannituna cabiasu   = the finger next the index finger.
9. bitin ötegla cabiasu   = hand next to complete.
10. catögladu   = 2 hands.

As if to emphasize the rarity of this method of forming numerals, the Jiviros afterward discarded the last five of the above scale, replacing them by words borrowed from the Quichuas, or ancient Peruvians. The same process may have been followed by other tribes, and in this way numerals which were originally digital may have disappeared. But we have no evidence that this has ever happened in any extensive manner. We are, rather, impelled to accept the occasional numerals of this class as exceptions to the general rule, until we have at our disposal further evidence of an exact and critical nature, which would cause us to modify this opinion. An elaborate philological study by Dr. J. H. Trumbull [86] of the numerals used by many of the North American Indian tribes reveals the presence in the languages of these tribes of a few, but only a few, finger names which are used without change as numeral expressions also. Sometimes the finger gives a name not its own to the numeral with which it is associated in counting—as in the Chippeway dialect, which has *nawi-nindj*, middle of the hand, and *nisswi*, 3; and the Cheyenne, where *notoyos*, middle finger, and *na-nohhtu*, 8, are closely related. In other parts of the world isolated examples of the transference of finger names to numerals are also found. Of these a well-known example is furnished by the Zulu numerals, where "*tatisitupa*, taking the thumb, becomes a numeral for six. Then the verb *komba*, to point, indicating the forefinger, or 'pointer,' makes the next numeral, seven. Thus, answering the question, 'How much did your master give you?' a Zulu would say, '*U kombile*,' 'He pointed with his forefinger,' *i.e.* 'He gave me seven'; and this curious way of using the numeral verb is also shown in such an example as '*amahasi akombile*,' 'the horses have pointed,' *i.e.* 'there were seven of them.' In like manner, *Kijangalobili*, 'keep back two fingers,' *i.e.* eight, and *Kijangalolunje*, 'keep back one finger,' *i.e.* nine, lead on to *kumi*, ten." [87]

Returning for a moment to the consideration of number systems in the formation of which the influence of the hand has been paramount, we find still further variations of the method already noticed of constructing names for the fives, tens, and twenties, as well as for the intermediate numbers. Instead of the simple words "hand," "foot," etc., we not infrequently meet with some paraphrase for one or for all these terms, the derivation of which is unmistakable. The Nengones, [88] an island tribe of the Indian Ocean, though using the word "man" for 20, do not employ explicit hand or foot words, but count

1.	sa.	
2.	rewe.	
3.	tini.	
4.	etse.	
5.	se dono	= the end (of the first hand).
6.	dono ne sa	= end and 1.
7.	dono ne rewe	= end and 2.
8.	dono ne tini	= end and 3.
9.	dono ne etse	= end and 4.
10.	rewe tubenine	= 2 series (of fingers).
11.	rewe tubenine ne sa re tsemene	= 2 series and 1 on the next?
20.	sa re nome	= 1 man.
30.	sa re nome ne rewe tubenine	= 1 man and 2 series.
40.	rewe ne nome	= 2 men.

Examples like the above are not infrequent. The Aztecs used for 10 the word *matlactli*, hand-half, *i.e.* the hand half of a man, and for

20 *cempoalli*, one counting. [89] The Point Barrow Eskimos call 10 *ko-dlin*, the upper part, *i.e.* of a man. One of the Ewe dialects of Western Africa [90] has *ewo*, done, for 10; while, curiously enough, 9, *asieke*, is a digital word, meaning "to part (from) the hand."

In numerous instances also some characteristic word not of hand derivation is found, like the Yoruba *ogodzi*, string, which becomes a numeral for 40, because 40 cowries made a "string"; and the Maori *tekau*, bunch, which signifies 10. The origin of this seems to have been the custom of counting yams and fish by "bunches" of ten each. [91]

Another method of forming numeral words above 5 or 10 is found in the presence of such expressions as second 1, second 2, etc. In languages of rude construction and incomplete development the simple numeral scale is often found to end with 5, and all succeeding numerals to be formed from the first 5. The progression from that point may be 5-1, 5-2, etc., as in the numerous quinary scales to be noticed later, or it may be second 1, second 2, etc., as in the Niam Niam dialect of Central Africa, where the scale is [92]

1.	sa.	
2.	uwi.	
3.	biata.	
4.	biama.	
5.	biswi.	
6.	batissa	= 2d 1.
7.	batiwwi	= 2d 2.
8.	batti-biata	= 2d 3.
9.	batti-biama	= 2d 4.
10.	bauwé	= 2d 5.

That this method of progression is not confined to the least developed languages, however, is shown by a most cursory examination of the numerals of our American Indian tribes, where numeral formation like that exhibited above is exceedingly common. In the Kootenay dialect, [93] of British Columbia, *qaetsa*, 4, and *wo-qaetsa*, 8, are obviously related, the latter word probably meaning a second 4.

Most of the native languages of British Columbia form their words for 7 and 8 from those which signify 2 and 3; as, for example, the Heiltsuk, [94] which shows in the following words a most obvious correspondence:

2.	matl.	7.	matlaaus.
3.	yutq.	8.	yutquaus.

In the Choctaw language [95] the relation between 2 and 7, and 3 and 8, is no less clear. Here the words are:

2.	tuklo.	7.	untuklo.
3.	tuchina.	8.	untuchina.

The Nez Percés [96] repeat the first three words of their scale in their 6, 7, and 8 respectively, as a comparison of these numerals will show.

1.	naks.	6.	oilaks.
2.	lapit.	7.	oinapt.
3.	mitat.	8.	oimatat.

In all these cases the essential point of the method is contained in the repetition, in one way or another, of the numerals of the second quinate, without the use with each one of the word for 5. This may make 6, 7, 8, and 9 appear as second 1, second 2, etc., or another 1, another 2, etc.; or, more simply still, as 1 more, 2 more, etc. It is the method which was briefly discussed in the early part of the present chapter, and is by no means uncommon. In a decimal scale this repetition would begin with 11 instead of 6; as in the system found in use in Tagala and Pampanaga, two of the Philippine Islands, where, for example, 11, 12, and 13 are: [97]

11.	labi-n-isa	= over 1.
12.	labi-n-dalaua	= over 2.
13.	labi-n-tatlo	= over 3.

A precisely similar method of numeral building is used by some of our Western Indian tribes. Selecting a few of the Assiniboine numerals [98] as an illustration, we have

11.	ak kai washe	= more 1.
12.	ak kai noom pah	= more 2.
13.	ak kai yam me nee	= more 3.
14.	ak kai to pah	= more 4.
15.	ak kai zap tah	= more 5.
16.	ak kai shak pah	= more 6, etc.

A still more primitive structure is shown in the numerals of the Mboushas [99] of Equatorial Africa. Instead of using 5-1, 5-2, 5-3, 5-4, or 2d 1, 2d 2, 2d 3, 2d 4, in forming their numerals from 6 to 9, they proceed in the following remarkable and, at first thought, inexplicable manner to form their compound numerals:

1.	ivoco.	
2.	beba.	
3.	belalo.	
4.	benai.	
5.	betano.	
6.	ivoco beba	= 1-2.
7.	ivoco belalo	= 1-3.
8.	ivoco benai	= 1-4.
9.	ivoco betano	= 1-5.
10.	dioum.	

No explanation is given by Mr. du Chaillu for such an apparently incomprehensible form of expression as, for example, 1-3, for 7. Some peculiar finger pantomime may accompany the counting, which, were it known, would enlighten us on the Mbousha's method of arriving at so anomalous a scale. Mere repetition in the second quinate of the words used in the first might readily be explained by supposing the use of fingers absolutely indispensable as an aid to counting, and that a certain word would have one meaning when associated with a certain finger of the left hand, and another meaning when associated with one of the fingers of the right. Such scales are, if the following are correct, actually in existence among the islands of the Pacific.

Balad. [100]

1.	parai.
2.	paroo.
3.	pargen.
4.	parbai.
5.	panim.
6.	parai.
7.	paroo.
8.	pargen.
9.	parbai.
10.	panim.

Uea. [100]

1.	tahi.
2.	lua.
3.	tolu.
4.	fa.
5.	lima.
6.	tahi.
7.	lua.
8.	tolu.
9.	fa.
10.	lima.

Such examples are, I believe, entirely unique among primitive number systems.

In numeral scales where the formative process has been of the general nature just exhibited, irregularities of various kinds are of frequent occurrence. Hand numerals may appear, and then suddenly disappear, just where we should look for them with the greatest degree of certainty. In the Ende, [101] a dialect of the Flores Islands, 5, 6, and 7 are of hand formation, while 8 and 9 are of entirely different origin, as the scale shows.

1.	sa.	
2.	zua.	
3.	telu.	
4.	wutu.	
5.	lima	
6.	lima sa	= hand 1.
7.	lima zua	= hand 2.
8.	rua butu	= 2 × 4.
9.	trasa	= 10 − 1?
10.	sabulu.	

One special point to be noticed in this scale is the irregularity that prevails between 7, 8, 9. The formation of 7 is of the most ordinary kind; 8 is 2 fours—common enough duplication; while 9 appears to be 10 − 1. All of these modes of compounding are, in their own way, regular; but the irregularity consists in using all three of them in connective numerals in the same system. But, odd as this jumble seems, it is more than matched by that found in the scale of the Karankawa Indians, [102] an extinct tribe formerly inhabiting the coast region of Texas. The first ten numerals of this singular array are:

1.	natsa.	
2.	haikia.	
3.	kachayi.	
4.	hayo hakn	= 2 × 2.
5.	natsa behema	= 1 father, *i.e.* of the fingers.
6.	hayo haikia	= 3 × 2?
7.	haikia natsa	= 2 + 5?
8.	haikia behema	= 2 fathers?
9.	haikia doatn	= 2d from 10?
10.	doatn habe.	

Systems like the above, where chaos instead of order seems to be the ruling principle, are of occasional occurrence, but they are decidedly the exception.

In some of the cases that have been adduced for illustration it is to be noticed that the process of combination begins with 7 instead of with 6. Among others, the scale of the Pigmies of Central Africa [103] and that of the Mosquitos [104] of Central America show this tendency. In the Pigmy scale the words for 1 and 6 are so closely akin that one cannot resist the impression that 6 was to them a new 1, and was thus named.

	Mosquito.	**Pigmy.**
1.	kumi.	ujju.
2.	wal.	ibari.
3.	niupa.	ikaro.
4.	wal-wal = 2-2.	ikwanganya.
5.	mata-sip = fingers of 1 hand.	bumuti.
6.	matlalkabe.	ijju.
7.	matlalkabe pura kumi = 6 and 1.	bumutti-na-ibali = 5 and 2.
8.	matlalkabe pura wal = 6 and 2.	bumutti-na-ikaro = 5 and 3.
9.	matlalkabe pura niupa = 6 and 3.	bumutti-na-ikwanganya = 5 and 4.
10.	mata wal sip = fingers of 2 hands.	mabo = half man.

The Mosquito scale is quite exceptional in forming 7, 8, and 9 from 6, instead of from 5. The usual method, where combinations appear between 6 and 10, is exhibited by the Pigmy scale. Still another species of numeral form, quite different from any that have already been noticed, is found in the Yoruba [105] scale, which is in many respects one of the most peculiar in existence. Here the words for 11, 12, etc., are formed by adding the suffix *-la*, great, to the words for 1, 2, etc., thus:

1.	eni, or okan.
2.	edzi.
3.	eta.
4.	erin.

5.	arun.	
6.	efa.	
7.	edze.	
8.	edzo.	
9.	esan.	
10.	ewa.	
11.	okanla	= great 1.
12.	edzila	= great 2.
13.	etala	= great 3.
14.	erinla	= great 4, etc.
40.	ogodzi	= string.
200.	igba	= heap.

The word for 40 was adopted because cowrie shells, which are used for counting, were strung by forties; and *igba*, 200, because a heap of 200 shells was five strings, and thus formed a convenient higher unit for reckoning. Proceeding in this curious manner, [106] they called 50 strings 1 *afo* or head; and to illustrate their singular mode of reckoning—the king of the Dahomans, having made war on the Yorubans, and attacked their army, was repulsed and defeated with a loss of "two heads, twenty strings, and twenty cowries" of men, or 4820.

The number scale of the Abipones, [107] one of the low tribes of the Paraguay region, contains two genuine curiosities, and by reason of those it deserves a place among any collection of numeral scales designed to exhibit the formation of this class of words. It is:

1.	initara	= 1 alone.
2.	inoaka.	
3.	inoaka yekaini	= 2 and 1.
4.	geyenknate	= toes of an ostrich.
5.	neenhalek	= a five coloured, spotted hide,
	or hanambegen	= fingers of 1 hand.
10.	lanamrihegem	= fingers of both hands.

20. lanamrihegem cat gracherhaka anamichirihegem = fingers of both hands together with toes of both feet.

That the number sense of the Abipones is but little, if at all, above that of the native Australian tribes, is shown by their expressing 3 by the combination 2 and 1. This limitation, as we have already seen, is shared by the Botocudos, the Chiquitos, and many of the other native races of South America. But the Abipones, in seeking for words with which to enable themselves to pass beyond the limit 3, invented the singular terms just given for 4 and 5. The ostrich, having three toes in front and one behind on each foot presented them with a living example of 3 + 1; hence "toes of an ostrich" became their numeral for 4. Similarly, the number of colours in a certain hide being five, the name for that hide was adopted as their next numeral. At this point they began to resort to digital numeration also; and any higher number is expressed by that method.

In the sense in which the word is defined by mathematicians, *number* is a pure, abstract concept. But a moment's reflection will show that, as it originates among savage races, number is, and from the limitations of their intellect must be, entirely concrete. An abstract conception is something quite foreign to the essentially primitive mind, as missionaries and explorers have found to their chagrin. The savage can form no mental concept of what civilized man means by such a word as "soul"; nor would his idea of the abstract number 5 be much clearer. When he says *five*, he uses, in many cases at least, the same word that serves him when he wishes to say *hand*; and his mental concept when he says *five* is of a hand. The concrete idea of a closed fist or an open hand with outstretched fingers, is what is upper-most in his mind. He knows no more and cares no more about the pure number 5 than he does about the law of the conservation of energy. He sees in his mental picture only the real, material image, and his only comprehension of the number is, "these objects are as many as the fingers on my hand." Then, in the lapse of the long interval of centuries which intervene between lowest barbarism and highest civilization, the abstract and the concrete become slowly dissociated, the one from the other. First the actual hand picture fades away, and the number is recognized without the original assistance furnished by the derivation of the word. But the number is still for a long time a certain number *of objects*, and not an independent concept. It is only when the savage ceases to be wholly an animal, and becomes a thinking human being, that number in the abstract can come within the grasp of his mind. It is at this point that mere reckoning ceases, and arithmetic begins.

Chapter IV.

The Origin of Number Words.
(*Continued.*)

By the slow, and often painful, process incident to the extension and development of any mental conception in a mind wholly un-used to abstractions, the savage gropes his way onward in his counting from 1, or more probably from 2, to the various higher numbers required to form his scale. The perception of unity offers no difficulty to his mind, though he is conscious at first of the object itself rather than of any idea of number associated with it. The con-cept of duality, also, is grasped with perfect readiness. This concept is, in its simplest form, presented to the mind as soon as the indi-vidual distinguishes himself from another person, though the idea is still essentially concrete. Perhaps the first glimmering of any real number thought in connection with 2 comes when the savage con-trasts one single object with another — or, in other words, when he first recognizes the *pair*. At first the individuals composing the pair are simply "this one," and "that one," or "this and that"; and his number system now halts for a time at the stage when he can, rude-ly enough it may be, count 1, 2, many. There are certain cases where the forms of 1 and 2 are so similar ~~than~~ that one may readily imag-ine that these numbers really were "this" and "that" in the savage's original conception of them; and the same likeness also occurs in the words for 3 and 4, which may readily enough have been a second "this" and a second "that." In the Lushu tongue the words for 1 and 2 are *tizi* and *tazi* respectively. In Koriak we find *ngroka*, 3, and *ngraka*, 4; in Kolyma, *niyokh*, 3, and *niyakh*, 4; and in Kamtschatkan, *tsuk*, 3, and *tsaak*, 4. [108] Sometimes, as in the case of the Australian races, the entire extent of the count is carried through by means of pairs. But the natural theory one would form is, that 2 is the halting place for a very long time; that up to this point the fingers may or may not have been used — probably not; and that when the next start is made, and 3, 4, 5, and so on are counted, the fingers first come into requisition. If the grammatical structure of the earlier languages of the world's history is examined, the student is struck with the prevalence of the dual number in them—something which tends to disappear as language undergoes extended development. The dual number points unequivocally to the time when 1 and 2 were *the* numbers at mankind's disposal; to the time when his three numeral concepts, 1, 2, many, each demanded distinct expression. With increasing knowledge the necessity for this differentiatuin would pass away, and but two numbers, singular and plural, would remain. Incidentally it is to be noticed that the Indo-European words for 3—*three, trois, drei, tres, tri,* etc., have the same root as the Latin *trans,* beyond, and give us a hint of the time when our Aryan ancestors counted in the manner I have just described.

The first real difficulty which the savage experiences in counting, the difficulty which comes when he attempts to pass beyond 2, and to count 3, 4, and 5, is of course but slight; and these numbers are commonly used and readily understood by almost all tribes, no matter how deeply sunk in barbarism we find them. But the instances that have already been cited must not be forgotten. The Chiquitos do not, in their primitive state, properly count at all; the Andamans, the Veddas, and many of the Australian tribes have no numerals higher than 2; others of the Australians and many of the South Americans stop with 3 or 4; and tribes which make 5 their limit are still more numerous. Hence it is safe to assert that even this insignificant number is not always reached with perfect ease. Beyond 5 primitive man often proceeds with the greatest difficulty. Most savages, even those of the tribes just mentioned, can really count above here, even though they have no words with which to express their thought. But they do it with reluctance, and as they go on they quickly lose all sense of accuracy. This has already been commented on, but to emphasize it afresh the well-known example given by Mr. Oldfield from his own experience among the Watchandies may be quoted. [109] "I once wished to ascertain the exact number of natives who had been slain on a certain occasion. The individual of whom I made the inquiry began to think over the names … assigning one of his fingers to each, and it was not until after many failures, and consequent fresh starts, that he was able to express so high a number, which he at length did by holding up his hand three times, thus giving me to understand that fifteen was the answer to this most difficult arithmetical question." This meagreness of knowledge in all things pertaining to numbers is often found to be sharply emphasized in the names adopted by savages for their numeral words. While discussing in a previous chapter the limits of number systems, we found many instances where anything above 2 or 3 was designated by some one of the comprehensive terms *much*, *many*, *very many*; these words, or such equivalents as *lot*, *heap*, or *plenty*, serving as an aid to the finger pantomime necessary to indicate numbers for which they have no real names. The low degree of intelligence and civilization revealed by such words is brought quite as sharply into prominence by the word occasionally found for 5. Whenever the fingers and hands are used at all, it would seem natural to expect for 5 some general expression signifying *hand*, for 10 *both hands*, and for 20 *man*. Such is, as we have already seen, the ordinary method of progression, but it is not universal. A drop in the scale of civilization takes us to a point where 10, instead of 20, becomes the whole man. The Kusaies, [110] of Strong's Island, call 10 *sie-nul*, 1 man, 30 *tol-nul*, 3 men, 40 *a naul*, 4 men, etc.; and the Ku-Mbutti [111] of central Africa have *mukko*, 10, and *moku*, man. If 10 is to be expressed by reference to the man, instead of his hands, it might appear more natural to employ some such expression as that adopted by the African Pigmies, [112] who call 10 *mabo*, and man *mabo-mabo*. With them, then, 10 is perhaps "half a man," as it actually is among the Towkas of South America; and we have already seen that with the Aztecs it was *matlactli*, the "hand half" of a man. [113] The same idea crops out in the expression used by the Nicobar Islanders for 30—*heam-umdjome ruktei*, 1 man (and a) half. [114] Such nomenclature is entirely natural, and it accords with the analogy offered by other words of frequent occurrence in the numeral scales of savage races. Still, to find 10 expressed by the term *man* always conveys an impression of mental poverty; though it may, of course, be urged that this might arise from the fact that some races never use the toes in counting, but go over the fingers again, or perhaps bring into requisition the fingers of a second man to express the second 10. It is not safe to postulate an extremely low degree of civilization from the presence of certain peculiarities of numeral formation. Only the most general statements can be ventured on, and these are always subject to modification through some circumstance connected with environment, mode of living, or intercourse with other tribes. Two

South American races may be cited, which seem in this respect to give unmistakable evidence of being sunk in deepest barbarism. These are the Juri and the Cayriri, who use the same word for man and for 5. The former express 5 by *ghomen apa*, 1 man, [115] and the latter by *ibicho*, person. [116] The Tasmanians of Oyster Bay use the native word of similar meaning, *puggana*, man, [117] for 5.

Wherever the numeral 20 is expressed by the term *man*, it may be expected that 40 will be 2 men, 60, 3 men, etc. This form of numeration is usually, though not always, carried as far as the system extends; and it sometimes leads to curious terms, of which a single illustration will suffice. The San Blas Indians, like almost all the other Central and South American tribes, count by digit numerals, and form their twenties as follows: [118]

20.	tula guena	= man 1.
40.	tula pogua	= man 2.
100.	tula atala	= man 5.
120.	tula nergua	= man 6.
1000.	tula wala guena	= great 1 man.

The last expression may, perhaps, be translated "great hundred," though the literal meaning is the one given. If 10, instead of 20, is expressed by the word "man," the multiples of 10 follow the law just given for multiples of 20. This is sufficiently indicated by the Kusaie scale; or equally well by the Api words for 100 and 200, which are [119]

> *duulimo toromomo* = 10 times the whole man.

> *duulimo toromomo va juo* = 10 times the whole man taken 2 times.

As an illustration of the legitimate result which is produced by the attempt to express high numbers in this manner the term applied by educated native Greenlanders [120] for a thousand may be cited. This numeral, which is, of course, not in common use, is

> *inuit kulit tatdlima nik kuleriartut navdlugit* = 10 men 5 times 10 times come to an end.

It is worth noting that the word "great," which appears in the scale of the San Blas Indians, is not infrequently made use of in the formation of higher numeral words. The African Mabas [121] call 10 *atuk*, great 1; the Hottentots [122] and the Hidatsa Indians call 100 great 10, their words being *gei disi* and *pitikitstia* respectively.

The Nicaraguans [123] express 100 by *guhamba*, great 10, and 400 by *dinoamba*, great 20; and our own familiar word "million," which so many modern languages have borrowed from the Italian, is nothing more nor less than a derivative of the Latin *mille*, and really means "great thousand." The Dakota [124] language shows the same origin for its expression of 1,000,000, which is *kick ta opong wa tunkah*, great 1000. The origin of such terms can hardly be ascribed to poverty of language. It is found, rather, in the mental association of the larger with the smaller unit, and the consequent repetition of the name of the smaller. Any unit, whether it be a single thing, a dozen, a score, a hundred, a thousand, or any other unit, is, whenever used, a single and complete group; and where the relation between them is sufficiently close, as in our "gross" and "great gross," this form of nomenclature is natural enough to render it a matter of some surprise that it has not been employed more frequently. An old English nursery rhyme makes use of this association, only in a manner precisely the reverse of that which appears now and then in numeral terms. In the latter case the process is always one of enlargement, and the associative word is "great." In the following rhyme, constructed by the mature for the amusement of the childish mind, the process is one of diminution, and the associative word is "little":

> One's none,
>
> Two's some,
>
> Three's a many,
>
> Four's a penny,
>
> Five's a little hundred. [125]

Any real numeral formation by the use of "little," with the name of some higher unit, would, of course, be impossible. The numeral scale must be complete before the nursery rhyme can be manufactured.

It is not to be supposed from the observations that have been made on the formation of savage numeral scales that all, or even the majority of tribes, proceed in the awkward and faltering manner indicated by many of the examples quoted. Some of the North American Indian tribes have numeral scales which are, as far as they go, as regular and almost as simple as our own. But where digital numeration is extensively resorted to, the expressions for higher numbers are likely to become complex, and to act as a real bar to the extension of the system. The same thing is true, to an even greater degree, of tribes whose number sense is so defective that they begin almost from the outset to use combinations. If a savage expresses the number 3 by the combination 2-1, it will at once be suspected that his numerals will, by the time he reaches 10 or 20, become so complex and confused that numbers as high as these will be expressed by finger pantomime rather than by words. Such is often the case; and the comment is frequently made by explorers that the tribes they have visited have no words for numbers higher than 3, 4, 5, 10, or 20, but that counting is carried beyond that point by the aid of fingers or other objects. So reluctant, in many cases, are savages to count by words, that limits have been assigned for spoken numerals, which subsequent investigation proved to fall far short of the real extent of the number systems to which they belonged. One of the south-western Indian tribes of the United States, the Comanches, was for a time supposed to have no numeral words below 10, but to count solely by the use of fingers. But the entire scale of this taciturn tribe was afterward discovered and published.

To illustrate the awkward and inconvenient forms of expression which abound in primitive numeral nomenclature, one has only to draw from such scales as those of the Zuñi, or the Point Barrow Eskimos, given in the last chapter. Terms such as are found there may readily be duplicated from almost any quarter of the globe. The Soussous of Sierra Leone [126] call 99 *tongo solo manani nun solo manani,* *i.e.* to take (10 understood) 5 + 4 times and 5 + 4. The Malagasy expression for 1832 is [127] *roambistelo polo amby valonjato amby arivo,* 2 + 30 + 800 + 1000. The Aztec equivalent for 399 is [128] *caxtolli onnauh poalli ipan caxtolli onnaui,* (15 + 4) × 20 + 15 + 4; and the Sioux require for 29 the ponderous combination [129] *wick a chimen ne nompah sam pah nep e chu wink a.* These terms, long and awkward as they seem, are only the legitimate results which arise from combining the names of the higher and lower numbers, according to the peculiar genius of each language. From some of the Australian tribes are derived expressions still more complex, as for 6, *marh-jin-bang-ga-gudjir-gyn*, half the hands and 1; and for 15, *marh-jin-belli-belli-gudjir-jina-bang-ga*, the hand on either side and half the feet. [130] The Maré tribe, one of the numerous island tribes of Melanesia, [131] required for a translation of the numeral 38, which occurs in John v. 5, "had an infirmity thirty and eight years," the circumlocution, "one man and both sides five and three." Such expressions, curious as they seem at first thought, are no more than the natural outgrowth of systems built up by the slow and tedious process which so often obtains among primitive races, where digit numerals are combined in an almost endless variety of ways, and where mere reduplication often serves in place of any independent names for higher units. To what extent this may be carried is shown by the language of the Cayuba-bi, [132] who have for 10 the word *tunca*, and for 100 and 1000 the compounds *tunca tunca*, and *tunca tunca tunca* respectively; or of the Sapibocones, who call 10 *bururuche*, hand hand, and 100 *buruche buruche*, hand hand hand hand. [133] More remarkable still is the Ojibwa language, which continues its numeral scale without limit, furnishing combinations which are really remarkable; as, *e.g.*, that for 1,000,000,000, which is *me das wac me das wac as he me das wac*, [134] 1000 × 1000 × 1000. The Winnebago expression for the same number, [135] *ho ke he hhuta hhu chen a ho ke he ka ra pa ne za* is no less formidable, but it has every appearance of being an honest, native combination. All such primitive terms for larger numbers must, however, be received with caution. Savages are sometimes eager to display a knowledge they do not possess, and have been known to invent numeral words on the spot for the sake of carrying their scales to as high a limit as possible. The Choctaw words for million and billion are obvious attempts to incorporate the corresponding English terms into their own language. [136] For million they gave the vocabulary-hunter the phrase *mil yan chuffa*, and for billion, *bil yan chuffa.* The word *chuffa* signifies 1, hence these expressions are seen at a glance to be coined solely for the purpose of gratifying a little harmless Choctaw vanity. But this is innocence itself compared with the fraud perpetrated on Labillardière by the Tonga Islanders, who supplied the astonished and delighted investigator with a numeral vocabulary up to quadrillions. Their real limit was afterward found to be 100,000, and above that point they had palmed off as numerals a tolerably complete list of the obscene words of their language, together with a few nonsense terms. These were all accepted and printed in good faith, and the humiliating truth was not discovered until years afterward. [137]

One noteworthy and interesting fact relating to numeral nomenclature is the variation in form which words of this class undergo when applied to different classes of objects. To one accustomed as we are to absolute and unvarying forms for numerals, this seems at first a novel and almost unaccountable linguistic freak. But it is not uncommon among uncivilized races, and is extensively employed by so highly enlightened a people, even, as the Japanese. This variation in form is in no way analogous to that produced by inflectional changes, such as occur in Hebrew, Greek, Latin, etc. It is sufficient in many cases to produce almost an entire change in the form of the word; or to result in compounds which require close scrutiny for the detection of the original root. For example, in the Carrier, one of the Déné dialects of western Canada, the word *tha* means 3 things; *thane*, 3 persons; *that*, 3 times; *thatoen*, in 3 places; *thauh*, in 3 ways; *thailtoh*, all of the 3 things; *thahoeltoh*, all of the 3 persons; and *thahultoh*, all of the 3 times. [138] In the Tsimshian language of British Columbia we find seven distinct sets of numerals "which are used for various classes of objects that are counted. The first set is used in counting where there is no definite object referred to; the second class is used for counting flat objects and animals; the third for counting round objects and divisions of time; the fourth for counting men; the fifth for counting long objects, the numerals being composed with *kan*, tree; the sixth for counting canoes; and the seventh for measures. The last seem to be composed with *anon*, hand." [139] The first ten numerals of each of these classes is given in the following table:

No.	Counting	Flat Objects	Round Objects	Men	Long Objects	Canoes	Mea
1	gyak	gak	g'erel	k'al	k'awutskan	k'amaet	k'al

No.	Counting	Flat Objects	Round Objects	Men	Long Objects	Canoes	Meas...
2	t'epqat	t'epqat	goupel	t'epqadal	gaopskan	g'alpēeltk	gulbel
3	guant	guant	gutle	gulal	galtskan	galtskantk	guleon...
4	tqalpq	tqalpq	tqalpq	tqalpqdal	tqaapskan	tqalpqsk	tqalpqa...
5	kctōnc	kctōnc	kctōnc	kcenecal	k'etoentskan	kctōonsk	kctonsi...
6	k'alt	k'alt	k'alt	k'aldal	k'aoltskan	k'altk	k'aldel...
7	t'epqalt	t'epqalt	t'epqalt	t'epqaldal	t'epqaltskan	t'epqaltk	t'epqal...
8	guandalt	yuktalt	yuktalt	yuktleadal	ek'tlaedskan	yuktaltk	yuktal...
9	kctemac	kctemac	kctemac	kctemacal	kctemaestkan	kctemack	kctema...
10	gy'ap	gy'ap	kpēel	kpal	kpēetskan	gy'apsk	kpeont...

Remarkable as this list may appear, it is by no means as extensive as that derived from many of the other British Columbian tribes. The numerals of the Shushwap, Stlatlumh, Okanaken, and other languages of this region exist in several different forms, and can also be modified by any of the innumerable suffixes of these tongues. [140] To illustrate the almost illimitable number of sets that may be formed, a table is given of "a few classes, taken from the Heiltsuk dialect. [141] It appears from these examples that the number of classes is unlimited."

	One.	Two.	Three.
Animate.	menok	maalok	yutuk
Round.	menskam	masem	yutqsem
Long.	ments'ak	mats'ak	yututs'ak
Flat.	menaqsa	matlqsa	yutqsa
Day.	op'enequls	matlp'enequls	yutqp'enequls
Fathom.	op'enkh	matlp'enkh	yutqp'enkh
Grouped together.	– –	matloutl	yutoutl
Groups of objects.	nemtsmots'utl	matltsmots'utl	yutqtsmots'utl
Filled cup.	menqtlala	matl'aqtlala	yutqtlala
Empty cup.	menqtla	matl'aqtla	yutqtla
Full box.	menskamala	masemala	yutqsemala

	One.	**Two.**	**Three.**
Empty box.	menskam	masem	yutqsem
Loaded canoe.	mentsake	mats'ake	yututs'ake
Canoe with crew.	ments'akis	mats'akla	yututs'akla
Together on beach.	— —	maalis	— —
Together in house, etc.	— —	maalitl	— —

Variation in numeral forms such as is exhibited in the above tables is not confined to any one quarter of the globe; but it is more universal among the British Columbian Indians than among any other race, and it is a more characteristic linguistic peculiarity of this than of any other region, either in the Old World or in the New. It was to some extent employed by the Aztecs, [142] and its use is current among the Japanese; in whose language Crawfurd finds fourteen different classes of numerals "without exhausting the list." [143]

In examining the numerals of different languages it will be found that the tens of any ordinary decimal scale are formed in the same manner as in English. Twenty is simply 2 times 10; 30 is 3 times 10, and so on. The word "times" is, of course, not expressed, any more than in English; but the expressions briefly are, 2 tens, 3 tens, etc. But a singular exception to this method is presented by the Hebrew, and other of the Semitic languages. In Hebrew the word for 20 is the plural of the word for 10; and 30, 40, 50, etc. to 90 are plurals of 3, 4, 5, 6, 7, 8, 9. These numerals are as follows: [144]

10,	eser,	20,	eserim,
3,	shalosh,	30,	shaloshim,
4,	arba,	40,	arbaim,
5,	chamesh,	50,	chamishshim,
6,	shesh,	60,	sheshshim,
7,	sheba,	70,	shibim,
8,	shemoneh,	80,	shemonim,
9,	tesha,	90,	tishim.

The same formation appears in the numerals of the ancient Phœnicians, [145] and seems, indeed, to be a well-marked characteristic of the various branches of this division of the Caucasian race. An analogous method appears in the formation of the tens in the Bisayan, [146] one of the Malay numeral scales, where 30, 40, … 90, are constructed from 3, 4, … 9, by adding the termination *-an*.

No more interesting contribution has ever been made to the literature of numeral nomenclature than that in which Dr. Trumbull embodies the results of his scholarly research among the languages of the native Indian tribes of this country. [147] As might be expected, we are everywhere confronted with a digital origin, direct or indirect, in the great body of the words examined. But it is clearly shown that such a derivation cannot be established for all numerals; and evidence collected by the most recent research fully substantiates the position taken by Dr. Trumbull. Nearly all the derivations established are such as to remind us of the meanings we have already seen recurring in one form or another in language after language. Five is the end of the finger count on one hand—as, the Micmac *nan*, and Mohegan *nunon*, gone, or spent; the Pawnee *sihuks*, hands half; the Dakota *zaptan*, hand turned down; and the Massachusetts *napanna*, on one side. Ten is the end of the finger count, but is not always expressed by the "both hands" formula so commonly met with. The Cree term for this number is *mitatat*, no further; and the corresponding word in Delaware is *m'tellen*, no more. The Dakota 10 is, like its 5, a straightening out of the fingers which have been turned over in counting, or *wickchemna*, spread out unbent. The same is true of the Hidatsa *pitika*, which signifies a smoothing out, or straightening. The Pawnee 4, *skitiks*, is unusual, signifying as it does "all the fingers," or more properly, "the fingers of the hand." The same meaning attaches to this numeral in a few other languages also, and reminds one of the habit some people have of beginning to count on the forefinger and proceeding from there to the little finger. Can this have been the habit of the tribes in question? A suggestion of the same nature is made by the Illinois and Miami words for 8, *parare* and *polane*, which signify "nearly ended." Six is almost always digital in origin, though the derivation may be indirect, as in the Illinois *kakatchui*, passing beyond the middle; and the Dakota *shakpe*, 1 in addition. Some of these significations are well matched by numerals from the Ewe scales of western Africa, where we find the following: [148]

1. de   = a going, *i.e.* a beginning. (Cf. the Zuñi *töpinte*, taken to start with.)

3. eto   = the father (from the middle, or longest finger).

6. ade   = the other going.

9. asieke = parting with the hands.

10. ewo   = done.

In studying the names for 2 we are at once led away from a strictly digital origin for the terms by which this number is expressed. These names seem to come from four different sources: (1) roots denoting separation or distinction; (2) likeness, equality, or opposition; (3) addition, *i.e.* putting to, or putting with; (4) coupling, pairing, or matching. They are often related to, and perhaps derived from, names of natural pairs, as feet, hands, eyes, arms, or wings. In the Dakota and Algonkin dialects 2 is almost always related to "arms" or "hands," and in the Athapaskan to "feet." But the relationship is that of common origin, rather than of derivation from these pair-names. In the Puri and Hottentot languages, 2 and "hand" are closely allied; while in Sanskrit, 2 may be expressed by any one of the words *kara*, hand, *bahu*, arm, *paksha*, wing, or *netra*, eye. [149] Still more remote from anything digital in their derivation are the following, taken at random from a very great number of examples that might be cited to illustrate this point. The Assiniboines call 7, *shak ko we*, or *u she nah*, the odd number. [150] The Crow 1, *hamat*, signifies "the least"; [151] the Mississaga 1, *pecik*, a very small thing. [152] In Javanese, Malay, and Manadu, the words for 1, which are respectively *siji*, *satu*, and *sabuah*, signify 1 seed, 1 pebble, and 1 fruit respectively [153] — words as natural and as much to be expected at the beginning of a number scale as any finger name could possibly be. Among almost all savage races one form or another of palpable arithmetic is found, such as counting by seeds, pebbles, shells, notches, or knots; and the derivation of number words from these sources can constitute no ground for surprise. The Marquesan word for 4 is *pona*, knot, from the practice of tying breadfruit in knots of 4. The Maori 10 is *tekau*, bunch, or parcel, from the counting of yams and fish by parcels of 10. [154] The Javanese call 25, *lawe*, a thread, or string; 50, *ekat*, a skein of thread; 400, *samas*, a bit of gold; 800, *domas*, 2 bits of gold. [155] The Macassar and Butong term for 100 is *bilangan*, 1 tale or reckoning. [156] The Aztec 20 is *cem pohualli*, 1 count; 400 is *centzontli*, 1 hair of the head; and 8000 is *xiquipilli*, sack. [157] This sack was of such a size as to contain 8000 cacao nibs, or grains, hence the derivation of the word in its numeral sense is perfectly natural. In Japanese we find a large number of terms which, as applied to the different units of the number scale, seem almost purely fanciful. These words, with their meanings as given by a Japanese lexicon, are as follows:

10,000, or 10^4,	män	= enormous number.
10^8,	oku	= a compound of the words "man" and "mind."
10^{12},	chio	= indication, or symptom.
10^{16},	kei	= capital city.
10^{20},	si	= a term referring to grains.
10^{24},	owi	= — —
10^{28},	jio	= extent of land.
10^{32},	ko	= canal.
10^{36},	kan	= some kind of a body of water.
10^{40},	sai	= justice.
10^{44},	sä	= support.
10^{48},	kioku	= limit, or more strictly, ultimate.
$.01^2$,	rin	= — —
$.01^3$,	mo	= hair (of some animal).
$.01^4$,	shi	= thread.

In addition to these, some of the lower fractional values are described by words meaning "very small," "very fine thread," "sand grain," "dust," and "very vague." Taken altogether, the Japanese number system is the most remarkable I have ever examined, in the extent and variety of the higher numerals with well-defined descriptive names. Most of the terms employed are such as to defy any attempt to trace the process of reasoning which led to their adoption. It is not improbable that the choice was, in some of these cases at least, either accidental or arbitrary; but still, the changes in word meanings which occur with the lapse of time may have differentiated significations originally alike, until no trace of kinship would appear to the casual observer. Our numerals "score" and "gross" are never thought of as having any original relation to what is conveyed by the other meanings which attach to these words. But the origin of each, which is easily traced, shows that, in the beginning, there existed a well-defined reason for the selection of these, rather than other terms, for the numbers they now describe. Possibly these remarkable Japanese terms may be accounted for in the same way, though the supposition is, for some reasons, quite improbable. The same may be said for the Malagasy 1000, *alina*, which also means "night," and the Hebrew 6, *shesh*, which has the additional signification "white marble," and the stray exceptions which now and then come to the light in this or that language. Such terms as these may admit of some logical explanation, but for the great mass of numerals whose primitive meanings can be traced at all, no explanation whatever is needed; the words are self-explanatory, as the examples already cited show.

A few additional examples of natural derivation may still further emphasize the point just discussed. In Bambarese the word for 10, *tank*, is derived directly from *adang*, to count. [158] In the language of Mota, one of the islands of Melanesia, 100 is *mel nol*, used and done with, referring to the leaves of the cycas tree, with which the count had been carried on. [159] In many other Melanesian dialects [160] 100 is *rau*, a branch or leaf. In the Torres Straits we find the same number expressed by *na won*, the close; and in Eromanga it is *narolim narolim* (2 × 5)(2 × 5). [161] This combination deserves remark only because of the involved form which seems to have been required for the expression of so small a number as 100. A compound instead of a simple term for any higher unit is never to be wondered at, so rude are some of the savage methods of expressing number; but "two fives (times) two fives" is certainly remarkable. Some form like that employed by the Nusqually [162] of Puget Sound for 1000, i.e. *padutssubquätche*, ten hundred, is more in accordance with primitive method. But we are equally likely to find such descriptive phrases for this numeral as the *dor paka*, banyan roots, of the Torres Islands; *rau na hai*, leaves of a tree, of Vaturana; or *udolu*, all, of the Fiji Islands. And two curious phrases for 1000 are those of the Banks' Islands, *tar mataqelaqela*, eye blind thousand, *i.e.* many beyond count; and of Malanta, *warehune huto*, opossum's hairs, or *idumie one*, count the sand. [163]

The native languages of India, Thibet, and portions of the Indian archipelago furnish us with abundant instances of the formation of secondary numeral scales, which were used only for special purposes, and without in any way interfering with the use of the number words already in use. "Thus the scholars of India, ages ago, selected a set of words for a memoria technica, in order to record dates and numbers. These words they chose for reasons which are still in great measure evident; thus 'moon' or 'earth' expressed 1, there being but one of each; 2 might be called 'eye,' 'wing,' 'arm,' 'jaw,' as going in pairs; for 3 they said 'Rama,' 'fire,' or 'quality,' there being considered to be three Ramas, three kinds of fire, three qualities (guna); for 4 were used 'veda,' 'age,' or 'ocean,' there being four of each recognized; 'season' for 6, because they reckoned six seasons; 'sage' or 'vowel,' for 7, from the seven sages and the seven vowels; and so on with higher numbers, 'sun' for 12, because of his twelve annual denominations, or 'zodiac' from his twelve signs, and 'nail' for 20, a word incidentally bringing in finger notation. As Sanskrit is very rich in synonyms, and as even the numerals themselves might be used, it became very easy to draw up phrases or nonsense verses to record series of numbers by this system of artificial memory." [164]

More than enough has been said to show how baseless is the claim that all numeral words are derived, either directly or indirectly, from the names of fingers, hands, or feet. Connected with the origin of each number word there may be some metaphor, which cannot always be distinctly traced; and where the metaphor was born of the hand or of the foot, we inevitably associate it with the practice of finger counting. But races as fond of metaphor and of linguistic embellishment as are those of the East, or as are our American Indians even, might readily resort to some other source than that furnished by the members of the human body, when in want of a term with which to describe the 5, 10, or any other number of the numeral scale they were unconsciously forming. That the first numbers of a numeral scale are usually derived from other sources, we have some reason to believe; but that all above 2, 3, or at most 4, are almost universally of digital origin we must admit. Exception should properly be made of higher units, say 1000 or anything greater, which could not be expected to conform to any law of derivation governing the first few units of a system.

Collecting together and comparing with one another the great mass of terms by which we find any number expressed in different languages, and, while admitting the great diversity of method practised by different tribes, we observe certain resemblances which were not at first supposed to exist. The various meanings of 1, where they can be traced at all, cluster into a little group of significations with which at last we come to associate the idea of unity. Similarly of 2, or 5, or 10, or any one of the little band which does picket duty for the advance guard of the great host of number words which are to follow. A careful examination of the first decade warrants the assertion that the probable meaning of any one of the units will be found in the list given below. The words selected are intended merely to serve as indications of the thought underlying the savage's choice, and not necessarily as the exact term by means of which he describes his number. Only the commonest meanings are included in the tabulation here given.

1   = existence, piece, group, beginning.
2   = repetition, division, natural pair.
3   = collection, many, two-one.
4   = two twos.
5   = hand, group, division,
6   = five-one, two threes, second one.
7   = five-two, second two, three from ten.
8   = five-three, second three, two fours, two from ten.
9   = five-four, three threes, one from ten.
10 = one (group), two fives (hands), half a man, one man.
15 = ten-five, one foot, three fives.
20 = two tens, one man, two feet. [165]

Chapter V.

Miscellaneous Number Bases.

In the development and extension of any series of numbers into a systematic arrangement to which the term *system* may be applied, the first and most indispensable step is the selection of some number which is to serve as a base. When the savage begins the process of counting he invents, one after another, names with which to designate the successive steps of his numerical journey. At first there is no attempt at definiteness in the description he gives of any considerable number. If he cannot show what he means by the use of his fingers, or perhaps by the fingers of a single hand, he unhesitatingly passes it by, calling it many, heap, innumerable, as many as the leaves on the trees, or something else equally expressive and equally indefinite. But the time comes at last when a greater degree of exactness is required. Perhaps the number 11 is to be indicated, and indicated precisely. A fresh mental effort is required of the ignorant child of nature; and the result is "all the fingers and one more," "both hands and one more," "one on another count," or some equivalent circumlocution. If he has an independent word for 10, the result will be simply ten-one. When this step has been taken, the base is established. The savage has, with entire unconsciousness, made all his subsequent progress dependent on the number 10, or, in other words, he has established 10 as the base of his number system. The process just indicated may be gone through with at 5, or at 20, thus giving us a quinary or a vigesimal, or, more probably, a mixed system; and, in rare instances, some other number may serve as the point of departure from simple into compound numeral terms. But the general idea is always the same, and only the details of formation are found to differ.

Without the establishment of some base any *system* of numbers is impossible. The savage has no means of keeping track of his count unless he can at each step refer himself to some well-defined milestone in his course. If, as has been pointed out in the foregoing chapters, confusion results whenever an attempt is made to count any number which carries him above 10, it must at once appear that progress beyond that point would be rendered many times more difficult if it were not for the fact that, at each new step, he has only to indicate the distance he has progressed beyond his base, and not the distance from his original starting-point. Some idea may, perhaps, be gained of the nature of this difficulty by imagining the numbers of our ordinary scale to be represented, each one by a single symbol different from that used to denote any other number. How long would it take the average intellect to master the first 50 even, so that each number could without hesitation be indicated by its appropriate symbol? After the first 50 were once mastered, what of the next 50? and the next? and the next? and so on. The acquisition of a scale for which we had no other means of expression than that just described would be a matter of the extremest difficulty, and could never, save in the most exceptional circumstances, progress beyond the attainment of a limit of a few hundred. If the various numbers in question were designated by words instead of by symbols, the difficulty of the task would be still further increased. Hence, the establishment of some number as a base is not only a matter of the very highest convenience, but of absolute necessity, if any save the first few numbers are ever to be used.

In the selection of a base, — of a number from which he makes a fresh start, and to which he refers the next steps in his count, — the savage simply follows nature when he chooses 10, or perhaps 5 or 20. But it is a matter of the greatest interest to find that other numbers have, in exceptional cases, been used for this purpose. Two centuries ago the distinguished philosopher and mathematician, Leibnitz, proposed a binary system of numeration. The only symbols needed in such a system would be 0 and 1. The number which is now symbolized by the figure 2 would be represented by 10; while 3, 4, 5, 6, 7, 8, etc., would appear in the binary notation as 11, 100, 101, 110, 111, 1000, etc. The difficulty with such a system is that it rapidly grows cumbersome, requiring the use of so many figures for indicating any number. But Leibnitz found in the representation of all numbers by means of the two digits 0 and 1 a fitting symbolization of the creation out of chaos, or nothing, of the entire universe by the power of the Deity. In commemoration of this invention a medal was struck bearing on the obverse the words

> Numero Deus impari gaudet,

and on the reverse,

> Omnibus ex nihilo ducendis sufficit Unum. [166]

This curious system seems to have been regarded with the greatest affection by its inventor, who used every endeavour in his power to bring it to the notice of scholars and to urge its claims. But it appears to have been received with entire indifference, and to have been regarded merely as a mathematical curiosity.

Unknown to Leibnitz, however, a binary method of counting actually existed during that age; and it is only at the present time that it is becoming extinct. In Australia, the continent that is unique in its flora, its fauna, and its general topography, we find also this anomaly among methods of counting. The natives, who are to be classed among the lowest and the least intelligent of the aboriginal races of the world, have number systems of the most rudimentary nature, and evince a decided tendency to count by twos. This peculiarity, which was to some extent shared by the Tasmanians, the island tribes of the Torres Straits, and other aboriginal races of that region, has by some writers been regarded as peculiar to their part of the world; as though a binary number system were not to be found elsewhere. This attempt to make out of the rude and unusual method of counting which obtained among the Australians a racial characteristic is hardly justified by fuller investigation. Binary number systems, which are given in full on another page, are found in South America. Some of the Dravidian scales are binary; [167] and the marked preference, not infrequently observed among savage races, for counting by pairs, is in itself a sufficient refutation of this theory. Still it is an unquestionable fact that this binary tendency is more pronounced among the Australians than among any other extensive number of kindred races. They seldom count in words above 4, and almost never as high as 7. One of the most careful observers among them expresses his doubt as to a native's ability to discover the loss of two pins, if he were first shown seven pins in a row, and then two were removed without his knowledge. [168] But he believes that if a single pin were removed from the seven, the Blackfellow would become conscious of its loss. This is due to his habit of counting by pairs, which enables him to discover whether any number within reasonable limit is odd or even. Some of the negro tribes of Africa, and of the Indian tribes of America, have the same habit. Progression by pairs may seem to some tribes as natural as progression by single units. It certainly is not at all rare; and in Australia its influence on spoken number systems is most apparent.

Any number system which passes the limit 10 is reasonably sure to have either a quinary, a decimal, or a vigesimal structure. A binary scale could, as it is developed in primitive languages, hardly extend to 20, or even to 10, without becoming exceedingly cumbersome. A binary scale inevitably suggests a wretchedly low degree of mental development, which stands in the way of the formation of any number scale worthy to be dignified by the name of system. Take, for example, one of the dialects found among the western tribes of the Torres Straits, where, in general, but two numerals are found to exist. In this dialect the method of counting is: [169]

1. urapun.
2. okosa.
3. okosa urapun   = 2-1.
4. okosa okosa   = 2-2.
5. okosa okosa urapun   = 2-2-1.
6. okosa okosa okosa   = 2-2-2.

Anything above 6 they call *ras*, a lot.

For the sake of uniformity we may speak of this as a "system." But in so doing, we give to the legitimate meaning of the word a severe strain. The customs and modes of life of these people are not such as to require the use of any save the scanty list of numbers given above; and their mental poverty prompts them to call 3, the first number above a single pair, 2-1. In the same way, 4 and 6 are respectively 2 pairs and 3 pairs, while 5 is 1 more than 2 pairs. Five objects, however, they sometimes denote by *urapuni-getal*, 1 hand. A precisely similar condition is found to prevail respecting the arithmetic of all the Australian tribes. In some cases only two numerals are found, and in others three. But in a very great number of the native languages of that continent the count proceeds by pairs, if indeed it proceeds at all. Hence we at once reject the theory that Australian arithmetic, or Australian counting, is essentially peculiar. It is simply a legitimate result, such as might be looked for in any part of the world, of the barbarism in which the races of that quarter of the world were sunk, and in which they were content to live.

The following examples of Australian and Tasmanian number systems show how scanty was the numerical ability possessed by these tribes, and illustrate fully their tendency to count by twos or pairs.

Murray River. [170]

1.	enea.	
2.	petcheval.	
3.	petchevalenea	= 2-1.
4.	petcheval peteheval	= 2-2.

Maroura.

1.	nukee.	
2.	barkolo.	
3.	barkolo nuke	= 2-1.
4.	barkolo barkolo	= 2-2.

Lake Kopperamana.

1.	ngerna.	
2.	mondroo.	
3.	barkooloo.	
4.	mondroo mondroo	= 2-2.

Mort Noular.

1.	gamboden.	
2.	bengeroo.	
3.	bengeroganmel	= 2-1.
4.	bengeroovor bengeroo	= 2 + 2.

Wimmera.

1.	keyap.	

2.	pollit.	
3.	pollit keyap	= 2-1.
4.	pollit pollit	= 2-2.

Popham Bay.

1.	motu.	
2.	lawitbari.	
3.	lawitbari-motu	= 2-1.

Kamilaroi. [171]

1.	mal.	
2.	bularr.	
3.	guliba.	
4.	bularrbularr	= 2-2.
5.	bulaguliba	= 2-3.
6.	gulibaguliba	= 3-3.

Port Essington. [172]

1.	erad.	
2.	nargarik.	
3.	nargarikelerad	= 2-1.
4.	nargariknargarik	= 2-2.

Warrego.

1.	tarlina.	
2.	barkalo.	
3.	tarlina barkalo	= 1-2.

Crocker Island.

1.	roka.	
2.	orialk.	
3.	orialkeraroka	= 2-1.

Warrior Island. [173]

| 1. | woorapoo. | |
| 2. | ocasara. | |

| 3. | ocasara woorapoo | = 2-1. |
| 4. | ocasara ocasara | = 2-2. |

Dippil. [174]

1. kalim.
2. buller.
3. boppa.
4. buller gira buller   = 2 + 2.
5. buller gira buller kalim   = 2 + 2 + 1.

Frazer's Island. [175]

1. kalim.
2. bulla.
3. goorbunda.
4. bulla-bulla   = 2-2.

Moreton's Bay. [176]

1. kunner.
2. budela.
3. muddan.
4. budela berdelu   = 2-2.

Encounter Bay. [177]

1. yamalaitye.
2. ningenk.
3. nepaldar.
4. kuko kuko   = 2-2, or pair pair.
5. kuko kuko ki   = 2-2-1.
6. kuko kuko kuko   = 2-2-2.
7. kuko kuko kuko ki   = 2-2-2-1.

Adelaide. [178]

1. kuma.
2. purlaitye, or bula.
3. marnkutye.

4.   yera-bula   = pair 2.
5.   yera-bula kuma   = pair 2-1.
6.   yera-bula purlaitye   = pair 2.2.

Wiraduroi. [179]

1.   numbai.
2.   bula.
3.   bula-numbai   = 2-1.
4.   bungu   = many.
5.   bungu-galan   = very many.

Wirri-Wirri. [180]

1.   mooray.
2.   boollar.
3.   belar mooray   = 2-1.
4.   boollar boollar   = 2-2.
5.   mongoonballa.
6.   mongun mongun.

Cooper's Creek. [181]

1.   goona.
2.   barkoola.
3.   barkoola goona   = 2-1.
4.   barkoola barkoola   = 2-2.

Bourke, Darling River. [182]

1.   neecha.
2.   boolla.
4.   boolla neecha   = 2-1.
3.   boolla boolla   = 2-2.

Murray River, N.W. Bend. [183]

1.   mata.
2.   rankool.
3.   rankool mata   = 2-1.

| 4. | rankool rankool | = 2-2. |

Yit-tha. [184]

1.	mo.	
2.	thral.	
3.	thral mo	= 2-1.
4.	thral thral	= 2-2.

Port Darwin. [185]

1.	kulagook.	
2.	kalletillick.	
3.	kalletillick kulagook	= 2-1.
4.	kalletillick kalletillick	= 2-2.

Champion Bay. [186]

1.	kootea.	
2.	woothera.	
3.	woothera kootea	= 2-1.
4.	woothera woothera	= 2-2.

Belyando River. [187]

1.	wogin.	
2.	booleroo.	
3.	booleroo wogin	= 2-1.
4.	booleroo booleroo	= 2-2.

Warrego River.

1.	onkera.	
2.	paulludy.	
3.	paulludy onkera	= 2-1.
4.	paulludy paulludy	= 2-2.

Richmond River.

1.	yabra.	
2.	booroora.	
3.	booroora yabra	= 2-1.

4.   booroora booroora   = 2-2.

Port Macquarie.

1.   warcol.
2.   blarvo.
3.   blarvo warcol   = 2-1.
4.   blarvo blarvo   = 2-2.

Hill End.

1.   miko.
2.   bullagut.
3.   bullagut miko   = 2-1.
4.   bullagut bullagut   = 2-2.

Moneroo

1.   boor.
2.   wajala, blala.
3.   blala boor   = 2-1.
4.   wajala wajala.

Gonn Station.

1.   karp.
2.   pellige.
3.   pellige karp   = 2-1.
4.   pellige pellige   = 2-2.

Upper Yarra.

1.   kaambo.
2.   benjero.
3.   benjero kaambo   = 2-2.
4.   benjero on benjero   = 2-2.

Omeo.

1.   bore.
2.   warkolala.
3.   warkolala bore   = 2-1.

4.   warkolala warkolala   = 2-2.

Snowy River.

1.   kootook.
2.   boolong.
3.   booloom catha kootook   = 2 + 1.
4.   booloom catha booloom   = 2 + 2.

Ngarrimowro.

1.   warrangen.
2.   platir.
3.   platir warrangen   = 2-1.
4.   platir platir   = 2-2.

This Australian list might be greatly extended, but the scales selected may be taken as representative examples of Australian binary scales. Nearly all of them show a structure too clearly marked to require comment. In a few cases, however, the systems are to be regarded rather as showing a trace of binary structure, than as perfect examples of counting by twos. Examples of this nature are especially numerous in Curr's extensive list — the most complete collection of Australian vocabularies ever made.

A few binary scales have been found in South America, but they show no important variation on the Australian systems cited above. The only ones I have been able to collect are the following:

Bakairi. [188]

1.   tokalole.
2.   asage.
3.   asage tokalo   = 2-1.
4.   asage asage   = 2-2.

Zapara. [189]

1. nuquaqui.
2. namisciniqui.
3. haimuckumarachi.
4. namisciniqui ckara maitacka   = 2 + 2.

5. namisciniqui ckara maitacka nuquaqui   = 2 pairs + 1.

6. haimuckumaracki ckaramsitacka   = 3 pairs.

Apinages. [190]

1.   pouchi.

2.   at croudou.

3.   at croudi-pshi   = 2-1.

4.   agontad-acroudo   = 2-2.

Cotoxo. [191]

1.   ihueto.

2.   ize.

3.   ize-te-hueto   = 2-1.

4.   ize-te-seze   = 2-2.

5.   ize-te-seze-hue   = 2-2-1.

Mbayi. [192]

1.   uninitegui.

2.   iniguata.

3.   iniguata dugani   = 2 over.

4.   iniguata driniguata   = 2-2.

5.   oguidi   = many.

Tama. [193]

1.   teyo.

2.   cayapa.

3.   cho-teyo   = 2 + 1.

4.   cayapa-ria   = 2 again.

5.   cia-jente   = hand.

Curetu. [194]

1.   tchudyu.

2.   ap-adyu.

3.   arayu.

4.   apaedyái   = 2 + 2.

5. 		tchumupa.

If the existence of number systems like the above are to be accounted for simply on the ground of low civilization, one might reasonably expect to find ternary and and quaternary scales, as well as binary. Such scales actually exist, though not in such numbers as the binary. An example of the former is the Betoya scale, [195] which runs thus:

1. edoyoyoi.
2. edoi 			= another.
3. ibutu 			= beyond.
4. ibutu-edoyoyoi 		= beyond 1, or 3-1.
5. ru-mocoso 		= hand.

The Kamilaroi scale, given as an example of binary formation, is partly ternary; and its word for 6, *guliba guliba*, 3-3, is purely ternary. An occasional ternary trace is also found in number systems otherwise decimal or quinary vigesimal; as the *dlkunoutl*, second 3, of the Haida Indians of British Columbia. The Karens of India [196] in a system otherwise strictly decimal, exhibit the following binary-ternary-quaternary vagary:

6. 	then tho 			$= 3 \times 2$.
7. 	then tho ta 		$= 3 \times 2\text{-}1$.
8. 	lwie tho 			$= 4 \times 2$.
9. 	lwie tho ta 		$= 4 \times 2\text{-}1$.

In the Wokka dialect, [197] found on the Burnett River, Australia, a single ternary numeral is found, thus:

1. 	karboon.
2. 	wombura.
3. 	chrommunda.
4. 	chrommuda karboon 					$= 3\text{-}1$.

Instances of quaternary numeration are less rare than are those of ternary, and there is reason to believe that this method of counting has been practised more extensively than any other, except the binary and the three natural methods, the quinary, the decimal, and the vigesimal. The number of fingers on one hand is, excluding the thumb, four. Possibly there have been tribes among which counting by fours arose as a legitimate, though unusual, result of finger counting; just as there are, now and then, individuals who count on their fingers with the forefinger as a starting-point. But no such practice has ever been observed among savages, and such theorizing is the merest guess-work. Still a definite tendency to count by fours is sometimes met with, whatever be its origin. Quaternary traces are repeatedly to be found among the Indian languages of British Columbia. In describing the Columbians, Bancroft says: "Systems of numeration are simple, proceeding by fours, fives, or tens, according to the different languages...." [198] The same preference for four is said to have existed in primitive times in the languages of Central Asia, and that this form of numeration, resulting in scores of 16 and 64, was a development of finger counting. [199]

In the Hawaiian and a few other languages of the islands of the central Pacific, where in general the number systems employed are decimal, we find a most interesting case of the development, within number scales already well established, of both binary and quaternary systems. Their origin seems to have been perfectly natural, but the systems themselves must have been perfected very slowly. In Tahitian, Rarotongan, Mangarevan, and other dialects found in the neighbouring islands of those southern latitudes, certain of the higher units, *tekau, rau, mano*, which originally signified 10, 100, 1000, have become doubled in value, and now stand for 20, 200, 2000. In Hawaiian and other dialects they have again been doubled, and there they stand for 40, 400, 4000. [200] In the Marquesas group both forms are found, the former in the southern, the latter in the northern, part of the archipelago; and it seems probable that one or both of these methods of numeration are scattered somewhat widely throughout that region. The origin of these methods is probably to be found in the fact that, after the migration from the west toward the east, nearly all the objects the natives would ever count in any great numbers were small,—as yams, cocoanuts, fish, etc.,—and would be most conveniently counted by pairs. Hence the native, as he counted one pair, two pairs, etc., might readily say *one, two*, and so on, omitting the word "pair" altogether. Having much more frequent occasion to employ this secondary than the primary meaning of his numerals, the native would easily allow the original significations to fall into disuse, and in the lapse of time to be entirely forgotten. With a subsequent migration to the northward a second duplication might take place, and so produce the singular effect of giving to the same numeral word three different meanings in different parts of Oceania. To illustrate the former or binary method of numeration, the Tahuatan, one of the southern dialects of the Marquesas group, may be employed. [201] Here the ordinary numerals are:

1.	tahi.
10.	onohuu.
20.	takau.
200.	au.
2,000.	mano.
20,000.	tini.
200,000.	tufa.
2,000,000.	pohi.

In counting fish, and all kinds of fruit, except breadfruit, the scale begins with *tauna*, pair, and then, omitting *onohuu*, they employ the same words again, but in a modified sense. *Takau* becomes 10, *au* 100, etc.; but as the word "pair" is understood in each case, the value is the same as before. The table formed on this basis would be:

2 (units)	= 1 tauna	= 2.
10 tauna	= 1 takau	= 20.
10 takau	= 1 au	= 200.
10 au	= 1 mano	= 2000.
10 mano	= 1 tini	= 20,000.
10 tini	= 1 tufa	= 200,000.
10 tufa	= 1 pohi	= 2,000,000.

For counting breadfruit they use *pona*, knot, as their unit, breadfruit usually being tied up in knots of four. *Takau* now takes its third signification, 40, and becomes the base of their breadfruit system, so to speak. For some unknown reason the next unit, 400, is expressed by *tauau*, while *au*, which is the term that would regularly stand for that number, has, by a second duplication, come to signify 800. The next unit, *mano*, has in a similar manner been twisted out of its original sense, and in counting breadfruit is made to serve for 8000. In the northern, or Nukuhivan Islands, the decimal-quaternary system is more regular. It is in the counting of breadfruit only, [202]

4 breadfruits	= 1 pona	= 4.
10 pona	= 1 toha	= 40.
10 toha	= 1 au	= 400.
10 au	= 1 mano	= 4000.
10 mano	= 1 tini	= 40,000.
10 tini	= 1 tufa	= 400,000.
10 tufa	= 1 pohi	= 4,000,000.

In the Hawaiian dialect this scale is, with slight modification, the universal scale, used not only in counting breadfruit, but any other objects as well. The result is a complete decimal-quaternary system, such as is found nowhere else in the world except in this and a few of the neighbouring dialects of the Pacific. This scale, which is almost identical with the Nukuhivan, is [203]

4 units	= 1 ha or tauna	= 4.
10 tauna	= 1 tanaha	= 40.
10 tanaha	= 1 lau	= 400.
10 lau	= 1 mano	= 4000.
10 mano	= 1 tini	= 40,000.
10 tini	= 1 lehu	= 400,000.

The quaternary element thus introduced has modified the entire structure of the Hawaiian number system. Fifty is *tanaha me ta umi*, 40 + 10; 76 is 40 + 20 + 10 + 6; 100 is *ua tanaha ma tekau*, 2 × 40 + 10; 200 is *lima tanaha*, 5 × 40; and 864,895 is 2 × 400,000 + 40,000 + 6 × 4000 + 2 × 400 + 2 × 40 + 10 + 5. [204] Such examples show that this secondary influence, entering and incorporating itself as a part of a well-developed decimal system, has radically changed it by the establishment of 4 as the primary number base. The role which 10 now plays is peculiar. In the natural formation of a quaternary scale new units would be introduced at 16, 64, 256, etc.; that is, at the square, the cube, and each successive power of the base. But, instead of this, the new units are introduced at 10 × 4, 100 × 4, 1000 × 4, etc.; that is, at the products of 4 by each successive power of the old base. This leaves the scale a decimal scale still, even while it may justly be called quaternary; and produces one of the most singular and interesting instances of number-system formation that has ever been observed. In this connection it is worth noting that these Pacific island number scales have been developed to very high limits — in some cases into the millions. The numerals for these large numbers do not seem in any way indefinite, but rather to convey to the mind of the native an idea as clear as can well be conveyed by numbers of such magnitude. Beyond the limits given, the islanders have indefinite expressions, but as far as can be ascertained these are only used when the limits given above have actually been passed. To quote one more example, the Hervey Islanders, who have a binary-decimal scale, count as follows:

5 kaviri (bunches of cocoanuts)	= 1 takau	= 20.
10 takau	= 1 rau	= 200.
10 rau	= 1 mano	= 2000.
10 mano	= 1 kiu	= 20,000.
10 kiu	= 1 tini	= 200,000.

Anything above this they speak of in an uncertain way, as *mano mano* or *tini tini*, which may, perhaps, be paralleled by our English phrases "myriads upon myriads," and "millions of millions." [205] It is most remarkable that the same quarter of the globe should present us with the stunted number sense of the Australians, and, side by side with it, so extended and intelligent an appreciation of numerical values as that possessed by many of the lesser tribes of Polynesia.

The Luli of Paraguay [206] show a decided preference for the base 4. This preference gives way only when they reach the number 10, which is an ordinary digit numeral. All numbers above that point belong rather to decimal than to quaternary numeration. Their numerals are:

1. alapea.
2. tamop.
3. tamlip.
4. lokep.
5. lokep moile alapea   = 4 with 1,
    or is-alapea   = hand 1.
6. lokep moile tamop   = 4 with 2.
7. lokep moile tamlip   = 4 with 3.
8. lokep moile lokep   = 4 with 4.
9. lokep moile lokep alapea   = 4 with 4-1.
10. is yaoum   = all the fingers of hand.
11. is yaoum moile alapea   = all the fingers of hand with 1.
20. is elu yaoum   = all the fingers of hand and foot.
30. is elu yaoum moile is-yaoum = all the fingers of hand and foot with all the fingers of hand.

Still another instance of quaternary counting, this time carrying with it a suggestion of binary influence, is furnished by the Mocobi [207] of the Parana region. Their scale is exceedingly rude, and they use the fingers and toes almost exclusively in counting; only using their spoken numerals when, for any reason, they wish to dispense with the aid of their hands and feet. Their first eight numerals are:

1. iniateda.
2. inabaca.
3. inabacao caini   = 2 above.
4. inabacao cainiba   = 2 above 2;
    or natolatata.
5. inibacao cainiba iniateda   = 2 above 2-1;
    or natolatata iniateda   = 4-1.
6. natolatatata inibaca   = 4-2.
7. natolata inibacao-caini   = 4-2 above.
8. natolata-natolata   = 4-4.

There is probably no recorded instance of a number system formed on 6, 7, 8, or 9 as a base. No natural reason exists for the choice of any of these numbers for such a purpose; and it is hardly conceivable that any race should proceed beyond the unintelligent binary or quaternary stage, and then begin the formation of a scale for counting with any other base than one of the three natural bases to which allusion has already been made. Now and then some anomalous fragment is found imbedded in an otherwise regular system, which carries us back to the time when the savage was groping his way onward in his attempt to give expression to some number greater than any he had ever used before; and now and then one of these fragments is such as to lead us to the border land of the might-have-been, and to cause us to speculate on the possibility of so great a numerical curiosity as a senary or a septenary scale. The Bretons call 18 *triouec'h*, 3-6, but otherwise their language contains no hint of counting by sixes; and we are left at perfect liberty to theorize at will on the existence of so unusual a number word. Pott remarks [208] that the Bolans, of western Africa, appear to make some use of 6 as their number base, but their system, taken as a whole, is really a quinary-decimal. The language of the Sundas, [209] or mountaineers of Java, contains traces of senary counting. The Akra words for 7 and 8, *paggu* and *paniu*, appear to mean 6-1 and 7-1, respectively; and the same is true of the corresponding Tambi words *pagu* and *panjo*. [210] The Watji tribe [211] call 6 *andee*, and 7 *anderee*, which probably means 6-1. These words are to be regarded as accidental variations on the ordinary laws of formation, and are no more significant of a desire to count by sixes than is the Wallachian term *deu-maw*, which expresses 18 as 2-9, indicates the existence of a scale of which 9 is the base. One remarkably interesting number system is that exhibited by the Mosquito tribe [212] of Central America, who possess an extensive quinary-vigesimal scale containing one binary and three senary compounds. The first ten words of this singular scale, which has already been quoted, are:

1. kumi.

2. wal.

3. niupa.

4. wal-wal   = 2-2.

5.   mata-sip   = fingers of one hand.

6.   matlalkabe.

7.   matlalkabe pura kumi   = 6 + 1.

8.   matlalkabe pura wal   = 6 + 2.

9.   matlalkabe pura niupa   = 6 + 3.

10. mata-wal-sip   = fingers of the second hand.

In passing from 6 to 7, this tribe, also, has varied the almost universal law of progression, and has called 7 6-1. Their 8 and 9 are formed in a similar manner; but at 10 the ordinary method is resumed, and is continued from that point onward. Few number systems contain as many as three numerals which are associated with 6 as their base. In nearly all instances we find such numerals singly, or at most in pairs; and in the structure of any system as a whole, they are of no importance whatever. For example, in the Pawnee, a pure decimal scale, we find the following odd sequence: [213]

6.   shekshabish.

7.   petkoshekshabish   = 2-6, *i.e.* 2d 6.

8.   touwetshabish   = 3-6, *i.e.* 3d 6.

9.   loksherewa   = 10 − 1.

In the Uainuma scale the expressions for 7 and 8 are obviously referred to 6, though the meaning of 7 is not given, and it is impossible to guess what it really does signify. The numerals in question are: [214]

6.   aira-ettagapi.

7.   aira-ettagapi-hairiwigani-apecapecapsi.

8.   aira-ettagapi-matschahma   = 6 + 2.

In the dialect of the Mille tribe a single trace of senary counting appears, as the numerals given below show: [215]

6.   dildjidji.

7.   dildjidji me djuun   = 6 + 1.

Finally, in the numerals used by the natives of the Marshall Islands, the following curiously irregular sequence also contains a single senary numeral: [216]

6.	thil thino	= 3 + 3.
7.	thilthilim-thuon	= 6 + 1.
8.	rua-li-dok	= 10 − 2.
9.	ruathim-thuon	= 10 − 2 + 1.

Many years ago a statement appeared which at once attracted attention and awakened curiosity. It was to the effect that the Maoris, the aboriginal inhabitants of New Zealand, used as the basis of their numeral system the number 11; and that the system was quite extensively developed, having simple words for 121 and 1331, *i.e.* for the square and cube of 11. No apparent reason existed for this anomaly, and the Maori scale was for a long time looked upon as something quite exceptional and outside all ordinary rules of number-system formation. But a closer and more accurate knowledge of the Maori language and customs served to correct the mistake, and to show that this system was a simple decimal system, and that the error arose from the following habit. Sometimes when counting a number of objects the Maoris would put aside 1 to represent each 10, and then those so set aside would afterward be counted to ascertain the number of tens in the heap. Early observers among this people, seeing them count 10 and then set aside 1, at the same time pronouncing the word *tekau*, imagined that this word meant 11, and that the ignorant savage was making use of this number as his base. This misconception found its way into the early New Zealand dictionary, but was corrected in later editions. It is here mentioned only because of the wide diffusion of the error, and the interest it has always excited. [217]

Aside from our common decimal scale, there exist in the English language other methods of counting, some of them formal enough to be dignified by the term *system* — as the sexagesimal method of measuring time and angular magnitude; and the duodecimal system of reckoning, so extensively used in buying and selling. Of these systems, other than decimal, two are noticed by Tylor, [218] and commented on at some length, as follows:

"One is the well-known dicing set, *ace, deuce, tray, cater, cinque, size*; thus *size-ace* is 6-1, *cinques* or *sinks*, double 5. These came to us from France, and correspond with the common French numerals, except *ace*, which is Latin *as*, a word of great philological interest, meaning 'one.' The other borrowed set is to be found in the *Slang Dictionary*. It appears that the English street-folk have adopted as a means of secret communication a set of Italian numerals from the organ-grinders and image-sellers, or by other ways through which Italian or Lingua Franca is brought into the low neighbourhoods of London. In so doing they have performed a philological operation not only curious but instructive. By copying such expressions as *due soldi, tre soldi*, as equivalent to 'twopence,' 'threepence,' the word *saltee* became a recognized slang term for 'penny'; and pence are reckoned as follows:

oney saltee	1d.	uno soldo.
dooe saltee	2d.	due soldi.
tray saltee	3d.	tre soldi.
quarterer saltee	4d.	quattro soldi.
chinker saltee	5d.	cinque soldi.
say saltee	6d.	sei soldi.
say oney saltee, or setter saltee	7d.	sette soldi.
say dooe saltee, or otter saltee	8d.	otto soldi.
say tray saltee, or nobba saltee	9d.	nove soldi.
say quarterer saltee, or dacha saltee	10d.	dieci soldi.
say chinker saltee or dacha oney saltee	11d.	undici soldi.
oney beong	1s.	
a beong say saltee	1s. 6d.	
dooe beong say saltee, or madza caroon	2s. 6d.	(half-crown, mezza corona).

One of these series simply adopts Italian numerals decimally. But the other, when it has reached 6, having had enough of novelty, makes 7 by 6-1, and so forth. It is for no abstract reason that 6 is thus made the turning-point, but simply because the costermonger is adding pence up to the silver sixpence, and then adding pence again up to the shilling. Thus our duodecimal coinage has led to the practice of counting by sixes, and produced a philological curiosity, a real senary notation."

In addition to the two methods of counting here alluded to, another may be mentioned, which is equally instructive as showing how readily any special method of reckoning may be developed out of the needs arising in connection with any special line of work. As is well known, it is the custom in ocean, lake, and river navigation to measure soundings by the fathom. On the Mississippi River, where constant vigilance is needed because of the rapid shifting of sand-bars, a special sounding nomenclature has come into vogue, [219] which the following terms will illustrate:

5	ft.	= five feet.
6	ft.	= six feet.
9	ft.	= nine feet.
10-1/2	ft.	= a quarter less twain; *i.e.* a quarter of a fathom less than 2.
12	ft.	= mark twain.
13-1/2	ft.	= a quarter twain.
16-1/2	ft.	= a quarter less three.
18	ft.	= mark three.
19-1/2	ft.	= a quarter three.
24	ft.	= deep four.

As the soundings are taken, the readings are called off in the manner indicated in the table; 10-1/2 feet being "a quarter less twain," 12 feet "mark twain," etc. Any sounding above "deep four" is reported as "no bottom." In the Atlantic and Gulf waters on the coast of this country the same system prevails, only it is extended to meet the requirements of the deeper soundings there found, and instead of "six feet," "mark twain," etc., we find the fuller expressions, "by the mark one," "by the mark two," and so on, as far as the depth requires. This example also suggests the older and far more widely diffused method of reckoning time at sea by bells; a system in which "one bell," "two bells," "three bells," etc., mark the passage of time for the sailor as distinctly as the hands of the clock could do it. Other examples of a similar nature will readily suggest themselves to the mind.

Two possible number systems that have, for purely theoretical reasons, attracted much attention, are the octonary and the duodecimal systems. In favour of the octonary system it is urged that 8 is an exact power of 2; or in other words, a large number of repeated halves can be taken with 8 as a starting-point, without producing a fractional result. With 8 as a base we should obtain by successive halvings, 4, 2, 1. A similar process in our decimal scale gives 5, 2-1/2, 1-1/4. All this is undeniably true, but, granting the argument up to this point, one is then tempted to ask "What of it?" A certain degree of simplicity would thereby be introduced into the Theory of Numbers; but the only persons sufficiently interested in this branch of mathematics to appreciate the benefit thus obtained are already trained mathematicians, who are concerned rather with the pure science involved, than with reckoning on any special base. A slightly increased simplicity would appear in the work of stockbrokers, and others who reckon extensively by quarters, eighths, and sixteenths. But such men experience no difficulty whatever in performing their mental computations in the decimal system; and they acquire through constant practice such quickness and accuracy of calculation, that it is difficult to see how octonary reckoning would materially assist them. Altogether, the reasons that have in the past been adduced in favour of this form of arithmetic seem trivial. There is no record of any tribe that ever counted by eights, nor is there the slightest likelihood that such a system could ever meet with any general favour. It is said that the ancient Saxons used the octonary system, [220] but how, or for what purposes, is not stated. It is not to be supposed that this was the common system of counting, for it is well known that the decimal scale was in use as far back as the evidence of language will take us. But the field of speculation into which one is led by the octonary scale has proved most attractive to some, and the conclusion has been soberly reached, that in the history of the Aryan race the octonary was to be regarded as the predecessor of the decimal scale. In support of this theory no direct evidence is brought forward, but certain verbal resemblances. Those *ignes fatuii* of the philologist are made to perform the duty of supporting an hypothesis which would never have existed but for their own treacherous suggestions. Here is one of the most attractive of them:

Between the Latin words *novus*, new, and *novem*, nine, there exists a resemblance so close that it may well be more than accidental. Nine is, then, the *new* number; that is, the first number on a new count, of which 8 must originally have been the base. Pursuing this thought by investigation into different languages, the same resemblance is found there. Hence the theory is strengthened by corroborative evidence. In language after language the same resemblance is found, until it seems impossible to doubt, that in prehistoric times, 9 *was* the new number—the beginning of a second tale. The following table will show how widely spread is this coincidence:

Sanskrit, navan	= 9.	nava	= new.
Persian, nuh	= 9.	nau	= new.
Greek, ἐννέα	= 9.	νέος	= new.
Latin, novem	= 9.	novus	= new.
German, neun	= 9.	neu	= new.
Swedish, nio	= 9.	ny	= new.
Dutch, negen	= 9.	nieuw	= new.
Danish, ni	= 9.	ny	= new.
Icelandic, nyr	= 9.	niu	= new.
English, nine	= 9.	new	= new.
French, neuf	= 9.	nouveau	= new.
Spanish, nueve	= 9.	neuvo	= new.
Italian, nove	= 9.	nuovo	= new.
Portuguese, nove	= 9.	novo	= new.
Irish, naoi	= 9.	nus	= new.
Welsh, naw	= 9.	newydd	= new.

Breton, nevez   = 9.   nuhue   = new. [221]

This table might be extended still further, but the above examples show how widely diffused throughout the Aryan languages is this resemblance. The list certainly is an impressive one, and the student is at first thought tempted to ask whether all these resemblances can possibly have been accidental. But a single consideration sweeps away the entire argument as though it were a cobweb. All the languages through which this verbal likeness runs are derived directly or indirectly from one common stock; and the common every-day words, "nine" and "new," have been transmitted from that primitive tongue into all these linguistic offspring with but little change. Not only are the two words in question akin in each individual language, but *they are akin in all the languages.* Hence all these resemblances reduce to a single resemblance, or perhaps identity, that between the Aryan words for "nine" and "new." This was probably an accidental resemblance, no more significant than any one of the scores of other similar cases occurring in every language. If there were any further evidence of the former existence of an Aryan octonary scale, the coincidence would possess a certain degree of significance; but not a shred has ever been produced which is worthy of consideration. If our remote ancestors ever counted by eights, we are entirely ignorant of the fact, and must remain so until much more is known of their language than scholars now have at their command. The word resemblances noted above are hardly more significant than those occurring in two Polynesian languages, the Fatuhivan and the Nakuhivan, [222] where "new" is associated with the number 7. In the former case 7 is *fitu,* and "new" is *fou;* in the latter 7 is *hitu,* and "new" is *hou.* But no one has, because of this likeness, ever suggested that these tribes ever counted by the senary method. Another equally trivial resemblance occurs in the Tawgy and the Kamassin languages, [223] thus:

Tawgy.

8.	siti-data	= 2 × 4.
9.	nameaitjuma	= another.

Kamassin.

8.	sin-the'de	= 2 × 4.

9.   amithun   = another.

But it would be childish to argue, from this fact alone, that either 4 or 8 was the number base used.

In a recent antiquarian work of considerable interest, the author examines into the question of a former octonary system of counting among the various races of the world, particularly those of Asia, and brings to light much curious and entertaining material respecting the use of this number. Its use and importance in China, India, and central Asia, as well as among some of the islands of the Pacific, and in Central America, leads him to the conclusion that there was a time, long before the beginning of recorded history, when 8 was the common number base of the world. But his conclusion has no basis in his own material even. The argument cannot be examined here, but any one who cares to investigate it can find there an excellent illustration of the fact that a pet theory may take complete possession of its originator, and reduce him finally to a state of infantile subjugation. [224]

Of all numbers upon which a system could be based, 12 seems to combine in itself the greatest number of advantages. It is capable of division by 2, 3, 4, and 6, and hence admits of the taking of halves, thirds, quarters, and sixths of itself without the introduction of fractions in the result. From a commercial stand-point this advantage is very great; so great that many have seriously advocated the entire abolition of the decimal scale, and the substitution of the duodecimal in its stead. It is said that Charles XII. of Sweden was actually contemplating such a change in his dominions at the time of his death. In pursuance of this idea, some writers have gone so far as to suggest symbols for 10 and 11, and to recast our entire numeral nomenclature to conform to the duodecimal base. [225] Were such a change made, we should express the first nine numbers as at present, 10 and 11 by new, single symbols, and 12 by 10. From this point the progression would be regular, as in the decimal scale — only the same combination of figures in the different scales would mean very different things. Thus, 17 in the decimal scale would become 15 in the duodecimal; 144 in the decimal would become 100 in the duodecimal; and 1728, the cube of the new base, would of course be represented by the figures 1000.

It is impossible that any such change can ever meet with general or even partial favour, so firmly has the decimal scale become intrenched in its position. But it is more than probable that a large part of the world of trade and commerce will continue to buy and sell by the dozen, the gross, or some multiple or fraction of the one or the other, as long as buying and selling shall continue. Such has been its custom for centuries, and such will doubtless be its custom for centuries to come. The duodecimal is not a natural scale in the same sense as are the quinary, the decimal, and the vigesimal; but it is a system which is called into being long after the complete development of one of the natural systems, solely because of the simple and familiar fractions into which its base is divided. It is the scale of civilization, just as the three common scales are the scales of nature. But an example of its use was long sought for in vain among the primitive races of the world. Humboldt, in commenting on the number systems of the various peoples he had visited during his travels, remarked that no race had ever used exclusively that best of bases, 12. But it has recently been announced [226] that the discovery of such a tribe had actually been made, and that the Aphos of Benuë, an African tribe, count to 12 by simple words, and then for 13 say 12-1, for 14, 12-2, etc. This report has yet to be verified, but if true it will constitute a most interesting addition to anthropological knowledge.

Chapter VI.

The Quinary System.

The origin of the quinary mode of counting has been discussed with some fulness in a preceding chapter, and upon that question but little more need be said. It is the first of the natural systems. When the savage has finished his count of the fingers of a single hand, he has reached this natural number base. At this point he ceases to use simple numbers, and begins the process of compounding. By some one of the numerous methods illustrated in earlier chapters, he passes from 5 to 10, using here the fingers of his second hand. He now has two fives; and, just as we say "twenty," *i.e.* two tens, he says "two hands," "the second hand finished," "all the fingers," "the fingers of both hands," "all the fingers come to an end," or, much more rarely, "one man." That is, he is, in one of the many ways at his command, saying "two fives." At 15 he has "three hands" or "one foot"; and at 20 he pauses with "four hands," "hands and feet," "both feet," "all the fingers of hands and feet," "hands and feet finished," or, more probably, "one man." All these modes of expression are strictly natural, and all have been found in the number scales which were, and in many cases still are, in daily use among the uncivilized races of mankind.

In its structure the quinary is the simplest, the most primitive, of the natural systems. Its base is almost always expressed by a word meaning "hand," or by some equivalent circumlocution, and its digital origin is usually traced without difficulty. A consistent formation would require the expression of 10 by some phrase meaning "two fives," 15 by "three fives," etc. Such a scale is the one obtained from the Betoya language, already mentioned in Chapter III., where the formation of the numerals is purely quinary, as the following indicate: [227]

5.	teente	= 1 hand.
10.	cayaente, or caya huena	= 2 hands.
15.	toazumba-ente	= 3 hands.

20.   caesa-ente   = 4 hands.

The same formation appears, with greater or less distinctness, in many of the quinary scales already quoted, and in many more of which mention might be made. Collecting the significant numerals from a few such scales, and tabulating them for the sake of convenience of comparison, we see this point clearly illustrated by the following:

Tamanac.

5.   amnaitone   = 1 hand.
10.   amna atse ponare   = 2 hands.

Arawak, Guiana.

5.   abba tekkabe   = 1 hand.
10.   biamantekkabe   = 2 hands.

Jiviro.

5.   alacötegladu   = 1 hand.
10.   catögladu   = 2 hands.

Niam Niam

5.   biswe
10.   bauwe   = 2d 5.

Nengones

5. se dono   = the end (of the fingers of 1 hand).
10. rewe tubenine   = 2 series (of fingers).

Sesake. [228]

5.   lima   = hand.
10.   dua lima   = 2 hands.

Ambrym. [229]

5.   lim   = hand.
10.   ra-lim   = 2 hands.

Pama. [229]

5.   e-lime   = hand.

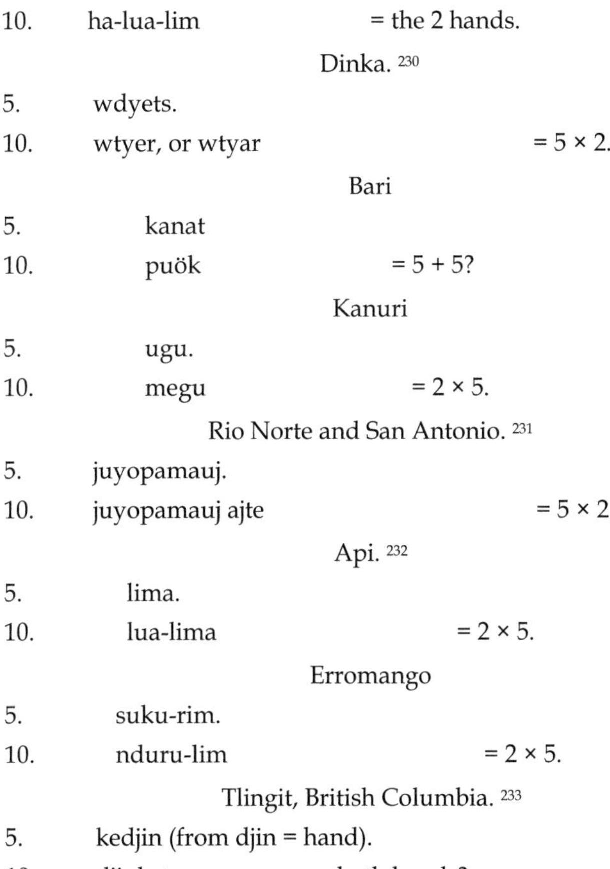

| 10. | ha-lua-lim | = the 2 hands. |

Dinka. [230]

| 5. | wdyets. | |
| 10. | wtyer, or wtyar | $= 5 \times 2.$ |

Bari

| 5. | kanat | |
| 10. | puök | $= 5 + 5?$ |

Kanuri

| 5. | ugu. | |
| 10. | megu | $= 2 \times 5.$ |

Rio Norte and San Antonio. [231]

| 5. | juyopamauj. | |
| 10. | juyopamauj ajte | $= 5 \times 2.$ |

Api. [232]

| 5. | lima. | |
| 10. | lua-lima | $= 2 \times 5.$ |

Erromango

| 5. | suku-rim. | |
| 10. | nduru-lim | $= 2 \times 5.$ |

Tlingit, British Columbia. [233]

| 5. | kedjin (from djin = hand). | |
| 10. | djinkat | = both hands? |

Thus far the quinary formation is simple and regular; and in view of the evidence with which these and similar illustrations furnish us, it is most surprising to find an eminent authority making the unequivocal statement that the number 10 is nowhere expressed by 2 fives [234] — that all tribes which begin their count on a quinary base express 10 by a simple word. It is a fact, as will be fully illustrated in the following pages, that quinary number systems, when extended, usually merge into either the decimal or the vigesimal. The result is, of course, a compound of two, and sometimes of three, systems in one scale. A pure quinary or vigesimal number system is exceedingly rare; but quinary scales certainly do exist in which, as far as we possess the numerals, no trace of any other influence appears. It is also to be noticed that some tribes, like the Eskimos of Point Barrow, though their systems may properly be classed as mixed systems, exhibit a decided preference for 5 as a base, and in counting objects, divided into groups of 5, obtaining the sum in this way. [235]

But the savage, after counting up to 10, often finds himself unconsciously impelled to depart from his strict reckoning by fives, and to assume a new basis of reference. Take, for example, the Zuñi system, in which the first 2 fives are:

5.	öpte	= the notched off.
10.	astem'thla	= all the fingers.

It will be noticed that the Zuñi does not say "two hands," or "the fingers of both hands," but simply "all the fingers." The 5 is no longer prominent, but instead the mere notion of one entire count of the fingers has taken its place. The division of the fingers into two sets of five each is still in his mind, but it is no longer the leading idea. As the count proceeds further, the quinary base may be retained, or it may be supplanted by a decimal or a vigesimal base. How readily the one or the other may predominate is seen by a glance at the following numerals:

Galibi. [236]

5.	atoneigne oietonaï	= 1 hand.
10.	oia batoue	= the other hand.
20.	poupoupatoret oupoume	= feet and hands.
40.	opoupoume	= twice the feet and hands.

Guarani. [237]

5.	ace popetei	= 1 hand.
10.	ace pomocoi	= 2 hands.
20.	acepo acepiabe	= hands and feet.

Fate. [238]

5.	lima	= hand.
10.	relima	= 2 hands.

20.   relima rua $= (2 × 5) × 2.$

Kiriri

5. mibika misa   = 1 hand.

10. mikriba misa sai   = both hands.

20. mikriba nusa ideko ibi sai   = both hands together with the feet.

Zamuco

5.   tsuena yimana-ite   = ended 1 hand.

10. tsuena yimana-die   = ended both hands.

20. tsuena yiri-die   = ended both feet.

Pikumbul

5.   mulanbu.

10. bularin murra   = belonging to the two hands.

15. mulanba dinna   = 5 toes added on (to the 10 fingers).

20. bularin dinna   = belonging to the 2 feet.

Yaruros. [239]

5.   kani-iktsi-mo   = 1 hand alone.

10.   yowa-iktsi-bo   = all the hands.

15.   kani-tao-mo   = 1 foot alone.

20.   kani-pume   = 1 man.

By the time 20 is reached the savage has probably allowed his conception of any aggregate to be so far modified that this number does not present itself to his mind as 4 fives. It may find expression in some phraseology such as the Kiriris employ—"both hands together with the feet"—or in the shorter "ended both feet" of the Zamucos, in which case we may presume that he is conscious that his count has been completed by means of the four sets of fives which are furnished by his hands and feet. But it is at least equally probable that he instinctively divides his total into 2 tens, and thus passes unconsciously from the quinary into the decimal scale. Again, the summing up of the 10 fingers and 10 toes often results in the concept of a single whole, a lump sum, so to speak, and the savage then says "one man," or something that gives utterance to this thought of a new unit. This leads the quinary into the vigesimal scale, and produces the combination so often found in certain parts of the world. Thus the inevitable tendency of any number system of quinary origin is toward the establishment of another and larger base, and the formation of a number system in which both are used. Wherever this is done, the greater of the two bases is always to be regarded as the principal number base of the language, and the 5 as entirely subordinate to it. It is hardly correct to say that, as a number system is extended, the quinary element disappears and gives place to the decimal or vigesimal, but rather that it becomes a factor of quite secondary importance in the development of the scale. If, for example, 8 is expressed by 5-3 in a quinary decimal system, 98 will be 9 × 10 + 5-3. The quinary element does not disappear, but merely sinks into a relatively unimportant position.

One of the purest examples of quinary numeration is that furnished by the Betoya scale, already given in full in Chapter III., and briefly mentioned at the beginning of this chapter. In the simplicity and regularity of its construction it is so noteworthy that it is worth repeating, as the first of the long list of quinary systems given in the following pages. No further comment is needed on it than that already made in connection with its digital significance. As far as given by Dr. Brinton the scale is:

1.	tey.	
2.	cayapa.	
3.	toazumba.	
4.	cajezea	= 2 with plural termination.
5.	teente	= hand.
6.	teyente tey	= hand 1.
7.	teyente cayapa	= hand 2.
8.	teyente toazumba	= hand 3.
9.	teyente caesea	= hand 4.
10.	caya ente, or caya huena	= 2 hands.
11.	caya ente-tey	= 2 hands 1.
15.	toazumba-ente	= 3 hands.
16.	toazumba-ente-tey	= 3 hands 1.

20. caesea ente   = 4 hands.

A far more common method of progression is furnished by languages which interrupt the quinary formation at 10, and express that number by a single word. Any scale in which this takes place can, from this point onward, be quinary only in the subordinate sense to which allusion has just been made. Examples of this are furnished in a more or less perfect manner by nearly all so-called quinary-vigesimal and quinary-decimal scales. As fairly representing this phase of number-system structure, I have selected the first 20 numerals from the following languages:

Welsh. [240]

1.	un.	
2.	dau.	
3.	tri.	
4.	pedwar.	
5.	pump.	
6.	chwech.	
7.	saith.	
8.	wyth.	
9.	naw.	
10.	deg.	
11.	un ar ddeg	= 1 + 10.
12.	deuddeg	= 2 + 10.
13.	tri ar ddeg	= 3 + 10.
14.	pedwar ar ddeg	= 4 + 10.
15.	pymtheg	= 5 + 10.
16.	un ar bymtheg	= 1 + 5 + 10.
17.	dau ar bymtheg	= 2 + 5 + 10.
18.	tri ar bymtheg	= 3 + 5 + 10.
19.	pedwar ar bymtheg	= 4 + 5 + 10.
20.	ugain.	

Nahuatl. [241]

1.	ce.	
2.	ome.	
3.	yei.	
4.	naui.	
5.	macuilli.	
6.	chiquacen	= [5] + 1.
7.	chicome	= [5] + 2.
8.	chicuey	= [5] + 3.
9.	chiucnaui	= [5] + 4.
10.	matlactli.	
11.	matlactli oce	= 10 + 1.
12.	matlactli omome	= 10 + 2.
13.	matlactli omey	= 10 + 3.
14.	matlactli onnaui	= 10 + 4.
15.	caxtolli.	
16.	caxtolli oce	= 15 + 1.
17.	caxtolli omome	= 15 + 2.
18.	caxtolli omey	= 15 + 3.
19.	caxtolli onnaui	= 15 + 4.
20.	cempualli	= 1 account.

Canaque [242] New Caledonia.

1.	chaguin.	
2.	carou.	
3.	careri.	
4.	caboue	
5.	cani.	
6.	cani-mon-chaguin	= 5 + 1.
7.	cani-mon-carou	= 5 + 2.
8.	cani-mon-careri	= 5 + 3.
9.	cani-mon-caboue	= 5 + 4.
10.	panrere.	

11.	panrere-mon-chaguin	= 10 + 1.
12.	panrere-mon-carou	= 10 + 2.
13.	panrere-mon-careri	= 10 + 3.
14.	panrere-mon-caboue	= 10 + 4.
15.	panrere-mon-cani	= 10 + 5.
16.	panrere-mon-cani-mon-chaguin	= 10 + 5 + 1.
17.	panrere-mon-cani-mon-carou	= 10 + 5 + 2.
18.	panrere-mon-cani-mon-careri	= 10 + 5 + 3.
19.	panrere-mon-cani-mon-caboue	= 10 + 5 + 4.
20.	jaquemo	= 1 person.

Guato. [243]

1.	cenai.	
2.	dououni.	
3.	coum.	
4.	dekai.	
5.	quinoui.	
6.	cenai-caicaira	= 1 on the other?
7.	dououni-caicaira	= 2 on the other?
8.	coum-caicaira	= 3 on the other?
9.	dekai-caicaira	= 4 on the other?
10.	quinoi-da	= 5 × 2.
11.	cenai-ai-caibo	= 1 + (the) hands.
12.	dououni-ai-caibo	= 2 + 10.
13.	coum-ai-caibo	= 3 + 10.
14.	dekai-ai-caibo	= 4 + 10.
15.	quin-oibo	= 5 × 3.
16.	cenai-ai-quacoibo	= 1 + 15.
17.	dououni-ai-quacoibo	= 2 + 15.
18.	coum-ai-quacoibo	= 3 + 15.
19.	dekai-ai-quacoibo	= 4 + 15.
20.	quinoui-ai-quacoibo	= 5 + 15.

The meanings assigned to the numerals 6 to 9 are entirely conjectural. They obviously mean 1, 2, 3, 4, taken a second time, and as the meanings I have given are often found in primitive systems, they have, at a venture, been given here.

Lifu, Loyalty Islands. [244]

1.	ca.	
2.	lue.	
3.	koeni.	
4.	eke.	
5.	tji pi.	
6.	ca ngemen	= 1 above.
7.	lue ngemen	= 2 above.
8.	koeni ngemen	= 3 above.
9.	eke ngemen	= 4 above.
10.	lue pi	= 2 × 5.
11.	ca ko.	
12.	lue ko.	
13.	koeni ko.	
14.	eke ko.	
15.	koeni pi	= 3 × 5.
16.	ca huai ano.	
17.	lua huai ano.	
18.	koeni huai ano.	
19.	eke huai ano.	
20.	ca atj	= 1 man.

Bongo. [245]

1. kotu.
2. ngorr.
3. motta.
4. neheo.
5. mui.

6. dokotu   = [5] + 1.

7. dongorr   = [5] + 2.

8. domotta   = [5] + 3.

9. doheo   = [5] + 4.

10. kih.

11. ki dokpo kotu   = 10 + 1.

12. ki dokpo ngorr   = 10 + 2.

13. ki dokpo motta   = 10 + 3.

14. ki dokpo neheo   = 10 + 4.

15. ki dokpo mui   = 10 + 5.

16. ki dokpo mui do mui okpo kotu   = 10 + 5 more, to 5, 1 more.

17. ki dokpo mui do mui okpo ngorr   = 10 + 5 more, to 5, 2 more.

18. ki dokpo mui do mui okpo motta   = 10 + 5 more, to 5, 3 more.

19. ki dokpo mui do mui okpo nehea   = 10 + 5 more, to 5, 4 more.

20. mbaba kotu.

Above 20, the Lufu and the Bongo systems are vigesimal, so that they are, as a whole, mixed systems.

The Welsh scale begins as though it were to present a pure decimal structure, and no hint of the quinary element appears until it has passed 15. The Nahuatl, on the other hand, counts from 5 to 10 by the ordinary quinary method, and then appears to pass into the decimal form. But when 16 is reached, we find the quinary influence still persistent; and from this point to 20, the numeral words in both scales are such as to show that the notion of counting by fives is quite as prominent as the notion of referring to 10 as a base. Above 20 the systems become vigesimal, with a quinary or decimal structure appearing in all numerals except multiples of 20. Thus, in Welsh, 36 is *unarbymtheg ar ugain*, 1 + 5 + 10 + 20; and in Nahuatl the same number is *cempualli caxtolli oce*, 20 + 15 + 1. Hence these and similar number systems, though commonly alluded to as vigesimal, are really mixed scales, with 20 as their primary base. The Canaque scale differs from the Nahuatl only in forming a compound word for 15, instead of introducing a new and simple term.

In the examples which follow, it is not thought best to extend the lists of numerals beyond 10, except in special instances where the illustration of some particular point may demand it. The usual quinary scale will be found, with a few exceptions like those just instanced, to have the following structure or one similar to it in all essential details: 1, 2, 3, 4, 5, 5-1, 5-2, 5-3, 5-4, 10, 10-1, 10-2, 10-3, 10-4, 10-5, 10-5-1, 10-5-2, 10-5-3, 10-5-4, 20. From these forms the entire system can readily be constructed as soon as it is known whether its principal base is to be 10 or 20.

Turning first to the native African languages, I have selected the following quinary scales from the abundant material that has been collected by the various explorers of the "Dark Continent." In some cases the numerals of certain tribes, as given by one writer, are found to differ widely from the same numerals as reported by another. No attempt has been made at comparison of these varying forms of orthography, which are usually to be ascribed to difference of nationality on the part of the collectors.

Feloops. [246]

1.	enory.	
2.	sickaba, or cookaba.	
3.	sisajee.	
4.	sibakeer.	
5.	footuck.	
6.	footuck-enory	= 5-1.
7.	footuck-cookaba	= 5-2.
8.	footuck-sisajee	= 5-3.
9.	footuck-sibakeer	= 5-4.
10.	sibankonyen.	

Kissi. [247]

1.	pili.
2.	miu.
3.	nga.
4.	iol.

5.	nguenu.	
6.	ngom-pum	= 5-1.
7.	ngom-miu	= 5-2.
8.	ngommag	= 5-3.
9.	nguenu-iol	= 5-4.
10.	to.	

Ashantee. [248]

1.	tah.	
2.	noo.	
3.	sah.	
4.	nah.	
5.	taw.	
6.	torata	= 5 + 1.
7.	toorifeenoo	= 5 + 2.
8.	toorifeessa	= 5 + 3.
9.	toorifeena	= 5 + 4.
10.	nopnoo.	

Basa. [249]

1.	do.	
2.	so.	
3.	ta.	
4.	hinye.	
5.	hum.	
6.	hum-le-do	= 5 + 1.
7.	hum-le-so	= 5 + 2.
8.	hum-le-ta	= 5 + 3.
9.	hum-le-hinyo	= 5 + 4.
10.	bla-bue.	

Jallonkas. [250]

| 1. | kidding. | |

2.	fidding.	
3.	sarra.	
4.	nani.	
5.	soolo.	
6.	seni.	
7.	soolo ma fidding	= 5 + 2.
8.	soolo ma sarra	= 5 + 3.
9.	soolo ma nani	= 5 + 4.
10.	nuff.	

Kru.

1.	da-do.	
2.	de-son.	
3.	de-tan.	
4.	de-nie.	
5.	de-mu.	
6.	dme-du	= 5-1.
7.	ne-son	= [5] + 2.
8.	ne-tan	= [5] + 3.
9.	sepadu	= 10 − 1?
10.	pua.	

Jaloffs. [251]

1.	wean.	
2.	yar.	
3.	yat.	
4.	yanet.	
5.	judom.	
6.	judom-wean	= 5-1.
7.	judom-yar	= 5-2.
8.	judom-yat	= 5-3.
9.	judom yanet	= 5-4.

10.   fook.

Golo. [252]

1.   mbali.
2.   bisi.
3.   bitta.
4.   banda.
5.   zonno.
6.   tsimmi tongbali $= 5 + 1.$
7.   tsimmi tobisi $= 5 + 2.$
8.   tsimmi tobitta $= 5 + 3.$
9.   tsimmi to banda $= 5 + 4.$
10.   nifo.

Foulah. [253]

1.   go.
2.   deeddee.
3.   tettee.
4.   nee.
5.   jouee.
6.   jego $= 5\text{-}1.$
7.   jedeeddee $= 5\text{-}2.$
8.   je-tettee $= 5\text{-}3.$
9.   je-nee $= 5\text{-}4.$
10.   sappo.

Soussou. [254]

1.   keren.
2.   firing.
3.   sarkan.
4.   nani.
5.   souli.
6.   seni.

7.	solo-fere	= 5-2.
8.	solo-mazarkan	= 5 + 3.
9.	solo-manani	= 5 + 4.
10.	fu.	

Bullom. [255]

1.	bul.	
2.	tin.	
3.	ra.	
4.	hyul.	
5.	men.	
6.	men-bul	= 5-1.
7.	men-tin	= 5-2.
8.	men-ra	= 5-3.
9.	men-hyul	= 5-4.
10.	won.	

Vei. [256]

1.	dondo.	
2.	fera.	
3.	sagba.	
4.	nani.	
5.	soru.	
6.	sun-dondo	= 5-1.
7.	sum-fera	= 5-2.
8.	sun-sagba	= 5-3.
9.	sun-nani	= 5-4.
10.	tan.	

Dinka. [257]

1.	tok.	
2.	rou.	
3.	dyak.	

4.	nuan.	
5.	wdyets.	
6.	wdetem	= 5-1.
7.	wderou	= 5-2.
8.	bet, bed	= 5-3.
9.	wdenuan	= 5-4.
10.	wtyer	= 5 × 2.

Temne.

1.	in.	
2.	ran.	
3.	sas.	
4.	anle.	
5.	tr-amat.	
6.	tr-amat rok-in	= 5 + 1.
7.	tr-amat de ran	= 5 + 2.
8.	tr-amat re sas	= 5 + 3.
9.	tr-amat ro n-anle	= 5 + 4.
10.	tr-ofatr.	

Abaker. [258]

1.	kili.	
2.	bore.	
3.	dotla.	
4.	ashe.	
5.	ini.	
6.	im kili	= 5-1.
7.	im-bone	= 5-2.
8.	ini-dotta	= 5-3.
9.	tin ashe	= 5-4.
10.	chica.	

Bagrimma. [259]

1.	kede.	
2.	sab.	
3.	muta.	
4.	so.	
5.	mi.	
6.	mi-ga	= 5 + 1.
7.	tsidi.	
8.	marta	= 5 + 2̶ 3̲.
9.	do-so	= [5] + 3̶ 4̲
10.	duk-keme.	

Papaa. [260]

1.	depoo.	
2.	auwi.	
3.	ottong.	
4.	enne.	
5.	attong.	
6.	attugo.	
7.	atjuwe	= [5] + 2.
8.	attiatong	= [5] + 3.
9.	atjeenne	= [5] + 4.
10.	awo.	

Efik. [261]

1.	kiet.	
2.	iba.	
3.	ita.	
4.	inan.	
5.	itiun.	
6.	itio-kiet	= 5-1.
7.	itia-ba	= 5-2.
8.	itia-eta	= 5-3.
9.	osu-kiet	= 10 − 1?

10.   duup.

Nupe. [262]

1.   nini.
2.   gu-ba.
3.   gu-ta.
4.   gu-ni.
5.   gu-tsun.
6.   gu-sua-yin   = 5 + 1.
7.   gu-tua-ba   = 5 + 2.
8.   gu-tu-ta   = 5 + 3.
9.   gu-tua-ni   = 5 + 4.
10.   gu-wo.

Mokko. [263]

1.   kiä.
2.   iba.
3.   itta.
4.   inan.
5.   üttin.
6.   itjüekee   = 5 + 1.
7.   ittiaba   = 5 + 2.
8.   itteiata   = 5 + 3.
9.   huschukiet.
10.   büb.

Kanuri. [264]

1.   tilo.
2.   ndi.
3.   yasge.
4.   dege.
5.   ugu.
6.   arasge   = 5 + 1.

7.	tulur.	
8.	wusge	= 5 + 3.
9.	legar.	
10.	megu	= 2 × 5.

Binin. [265]

1.	bo.	
2.	be.	
3.	la.	
4.	nin.	
5.	tang.	
6.	tahu	= 5 + 1?
7.	tabi	= 5 + 2.
8.	tara	= 5 + 3.
9.	ianin (tanin?)	= 5 + 4?
10.	te.	

Kredy. [266]

1.	baia.	
2.	rommu.	
3.	totto.	
4.	sosso.	
5.	saya.	
6.	yembobaia	= [5] + 1.
7.	yemborommu	= [5] + 2.
8.	yembototto	= [5] + 3.
9.	yembososso	= [5] + 4.
10.	puh.	

Herero. [267]

1.	mue.	
2.	vari.	
3.	tatu.	

4.   ne.

5.   tano.

6.   hambou-mue   = [5] + 1.

7.   hambou-vari   = [5] + 2.

8.   hambou-tatu   = [5] + 3.

9.   hambou-ne   = [5] + 4.

10.

Ki-Yau. [268]

1.   jumo.

2.   wawiri.

3.   watatu.

4.   mcheche.

5.   msano.

6.   musano na jumo   = 5 + 1.

7.   musano na wiri   = 5 + 2.

8.   musano na watatu   = 5 + 3.

9.   musano na mcheche   = 5 + 4.

10.   ikumi.

Fernando Po. [269]

1.   muli.

2.   mempa.

3.   meta.

4.   miene.

5.   mimito.

6.   mimito na muli   = 5 + 1.

7.   mimito na mempa   = 5 + 2.

8.   mimito na meta   = 5 + 3.

9.   mimito na miene   = 5 + 4.

10.   miemieu   = 5-5?

Ki-Nyassa

1.	kimodzi.	
2.	vi-wiri.	
3.	vi-tatu.	
4.	vinye.	
5.	visano.	
6.	visano na kimodzi	= 5 + 1.
7.	visano na vi-wiri	= 5 + 2.
8.	visano na vitatu	= 5 + 3.
9.	visano na vinye	= 5 + 4.
10.	chikumi.	

Balengue. [270]

1.	guevoho.	
2.	ibare.	
3.	raro.	
4.	inaï.	
5.	itano.	
6.	itano na guevoho	= 5 + 1.
7.	itano na ibare	= 5 + 2.
8.	itano na raro	= 5 + 3.
9.	itano na inaï	= 5 + 4.
10.	ndioum, or nai-hinaï.	

Kunama. [271]

1.	ella.	
2.	bare.	
3.	sadde.	
4.	salle.	
5.	kussume.	
6.	kon-t'-ella	= hand 1.
7.	kon-te-bare	= hand 2.
8.	kon-te-sadde	= hand 3.
9.	kon-te-salle	= hand 4.

10.   kol-lakada.

Gola. [272]

1.   ngoumou.
2.   ntie.
3.   ntaï.
4.   tina.
5.   nonon.
6.   diegoum   = [5] + 1.
7.   dientie   = [5] + 2.
8.   dietai   = [5] + 3.
9.   dectina   = [5] + 4.
10.   esia.

Barea. [273]

1.   doko
2.   arega.
3.   sane.
4.   sone.
5.   oita.
6.   data.
7.   dz-ariga   = 5 + 2.
8.   dis-sena   = 5 + 3.
9.   lefete-mada   = without 10.
10.   lefek.

Matibani. [274]

1.   mosa.
2.   pili.
3.   taru.
4.   teje.
5.   taru.
6.   tana mosa   = 5-1.

7.	tana pili	= 5-2.
8.	tana taru	= 5-3.
9.	loco.	
10.	loco nakege.	

Bonzé. [275]

1.	tan.	
2.	vele.	
3.	daba.	
4.	nani.	
5.	lolou.	
6.	maïda	= [5] + 1.
7.	maïfile	= [5] + 2.
8.	maïshaba	= [5] + 3.
9.	maïnan	= [5] + 4.
10.	bou.	

Mpovi

1.	moueta.	
2.	bevali.	
3.	betata.	
4.	benaï.	
5.	betani.	
6.	betani moueta	= 5-1.
7.	betani bevali	= 5-2.
8.	betani betata	= 5-3.
9.	betani benai	= 5-4.
10.	nchinia.	

Triton's Bay, New Quinea. [276]

1.	samosi.
2.	roueti.
3.	tourou.

4.   faat.

5.   rimi.

6.   rim-samosi   = 5-1.

7.   rim-roueti   = 5-2.

8.   rim-tourou   = 5-3.

9.   rim-faat   = 5-4.

10.   outsia.

Ende, or Flores. [277]

1.   sa.

2.   zua.

3.   telu.

4.   wutu.

5.   lima   = hand.

6.   lima-sa   = 5-1, or hand 1.

7.   lima-zua   = 5-2.

8.   rua-butu   = 2 × 4?

9.   trasa   = [10] − 1?

10.   sabulu.

Mallicolo. [278]

1.   tseekaee.

2.   ery.

3.   erei.

4.   ebats.

5.   ereem.

6.   tsookaee   = [5] + 1.

7.   gooy   = [5] + 2.

8.   hoorey   = [5] + 3.

9.   goodbats   = [5] + 4.

10.   senearn.

Ebon, Marshall Islands. [279]

1.	iuwun.	
2.	drud.	
3.	chilu.	
4.	emer.	
5.	lailem.	
6.	chilchinu	$= 5 + 1.$
7.	chilchime	$= 5 + 2.$
8.	twalithuk	$= [10] - 2.$
9.	twahmejuwou	$= [10] - 1.$
10.	iungou.	

Uea, Loyalty Island. [280]

1.	tahi.
2.	lua.
3.	tolu.
4.	fa.
5.	lima.
6.	tahi.
7.	lua.
8.	tolu.
9.	fa.
10.	lima.

Uea. [280] — [another dialect.]

1.	hacha.	
2.	lo.	
3.	kuun.	
4.	thack.	
5.	thabumb.	
6.	lo-acha	$= 2d\ 1.$
7.	lo-alo	$= 2d\ 2.$
8.	lo-kuun	$= 2d\ 3.$
9.	lo-thack	$= 2d\ 4.$

| 10. | lebenetee. | |

Isle of Pines. [281]

1.	ta.	
2.	bo.	
3.	beti.	
4.	beu.	
5.	ta-hue.	
6.	no-ta	= 2d 1.
7.	no-bo	= 2d 2.
8.	no-beti	= 2d 3.
9.	no-beu	= 2d 4.
10.	de-kau.	

Ureparapara, Banks Islands. [282]

1.	vo towa.	
2.	vo ro.	
3.	vo tol.	
4.	vo vet.	
5.	teveliem	= 1 hand.
6.	leve jea	= other 1.
7.	leve ro	= other 2.
8.	leve tol	= other 3.
9.	leve vet	= other 4.
10.	sanowul	= 2 sets.

Mota, Banks Islands. [282]

1.	tuwale.	
2.	nirua.	
3.	nitol.	
4.	nivat.	
5.	tavelima	= 1 hand.
6.	laveatea	= other 1.

7.	lavearua	= other 2.
8.	laveatol	= other 3.
9.	laveavat	= other 4.
10.	sanavul	= 2 sets.

New Caledonia. [283]

1.	parai.	
2.	paroo.	
3.	parghen.	
4.	parbai.	
5.	panim.	
6.	panim-gha	= 5-1.
7.	panim-roo	= 5-2.
8.	panim-ghen	= 5-3.
9.	panim-bai	= 5-4.
10.	parooneek.	

Yengen, New Cal. [284]

1.	hets.	
2.	heluk.	
3.	heyen.	
4.	pobits.	
5.	nim	= hand.
6.	nim-wet	= 5-1.
7.	nim-weluk	= 5-2.
8.	nim-weyen	= 5-3.
9.	nim-pobit	= 5-4.
10.	pain-duk.	

Aneiteum. [285]

1.	ethi.
2.	ero.
3.	eseik.

4.   manohwan.

5.   nikman.

6.   nikman cled et ethi   = 5 + 1.

7.   nikman cled et oro   = 5 + 2.

8.   nikman cled et eseik   = 5 + 3.

9.   nikman cled et manohwan   = 5 + 4.

10. nikman lep ikman   = 5 + 5.

Tanna

1.   riti.

2.   karu.

3.   kahar.

4.   kefa.

5.   krirum.

6.   krirum riti   = 5-1.

7.   krirum karu   = 5-2.

8.   krirum kahar?   = 5-3.

9.   krirum kefa?   = 5-4.

10. — —

Eromanga

1.   sai.

2.   duru.

3.   disil.

4.   divat.

5.   siklim   = 1 hand.

6.   misikai   = other 1?

7.   siklim naru   = 5-2.

8.   siklim disil   = 5-3.

9.   siklim mindivat   = 5 + 4.

10. narolim   = 2 hands.

Fate, New Heb. [286]

1.	iskei.	
2.	rua.	
3.	tolu.	
4.	bate.	
5.	lima	= hand.
6.	la tesa	= other 1.
7.	la rua	= other 2.
8.	la tolu	= other 3.
9.	la fiti	= other 4.
10.	relima	= 2 hands.

Api, New Heb.

1.	tai.	
2.	lua.	
3.	tolu.	
4.	vari.	
5.	lima	= hand.
6.	o rai	= other 1.
7.	o lua	= other 2.
8.	o tolo	= other 3.
9.	o vari	= other 4.
10.	lua lima	= 2 hands.

Sesake, New Heb.

1.	sikai.	
2.	dua.	
3.	dolu.	
4.	pati.	
5.	lima	= hand.
6.	la tesa	= other 1.
7.	la dua	= other 2.
8.	la dolu	= other 3.
9.	lo veti	= other 4.

| 10. | dua lima | = 2 hands. |

Pama, New Heb.

1.	tai.	
2.	e lua.	
3.	e tolu.	
4.	e hati.	
5.	e lime	= hand.
6.	a hitai	= other 1.
7.	o lu	= other 2.
8.	o tolu	= other 3.
9.	o hati	= other 4.
10.	ha lua lim	= 2 hands

Aurora, New Heb.

1.	tewa.	
2.	i rua.	
3.	i tol.	
4.	i vat.	
5.	tavalima	= 1 hand.
6.	lava tea	= other 1.
7.	lava rua	= other 2.
8.	lava tol	= other 3.
9.	la vat	= other 4.
10.	sanwulu	= two sets.

Tobi. [287]

1.	yat.	
2.	glu.	
3.	ya.	
4.	uan.	
5.	yanim	= 1 hand.
6.	yawor	= other 1.

7.	yavic	= other 2.
8.	yawa	= other 3.
9.	yatu	= other 4.
10.	yasec.	

Palm Island. [288]

1.	yonkol.	
2.	yakka.	
3.	tetjora.	
4.	tarko.	
5.	yonkol mala	= 1 hand.

Jajowerong, Victoria. [288]

1.	kiarp.	
2.	bulaits.	
3.	bulaits kiarp	= 2-1.
4.	bulaits bulaits	= 2-2.
5.	kiarp munnar	= 1 hand.
6.	bulaits bulaits bulaits	= 2-2-2.
10.	bulaits munnar	= 2 hands.

The last two scales deserve special notice. They are Australian scales, and the former is strongly binary, as are so many others of that continent. But both show an incipient quinary tendency in their names for 5 and 10.

Cambodia. [289]

1.	muy.	
2.	pir.	
3.	bey.	
4.	buon.	
5.	pram.	
6.	pram muy	= 5-1.
7.	pram pil	= 5-2.
8.	pram bey	= 5-3.

9.   pram buon   = 5-4.

10.   dap.

Tschukschi. [290]

1.   inen.

2.   nirach.

3.   n'roch.

4.   n'rach.

5.   miligen   = hand.

6.   inen miligen   = 1-5.

7.   nirach miligen   = 2-5.

8.   anwrotkin.

9.   chona tsinki.

10.   migitken   = both hands.

Kottisch [291]

1.   hutsa.

2.   ina.

3.   tona.

4.   sega.

5.   chega.

6.   chelutsa   = 5 + 1.

7.   chelina   = 5 + 2.

8.   chaltona   = 5 + 3.

9.   tsumnaga   = 10 − 1.

10.   haga.

Eskimo of N.-W. Alaska. [292]

1.   a towshek.

2.   hipah, or malho.

3.   pingishute.

4.   sesaimat.

5.   talema.

6.   okvinile, or ahchegaret   = another 1?
7.   talema-malronik   = 5-two of them.
8.   pingishu-okvingile   = 2d 3?
9.   kolingotalia   = 10 − 1?
10.   koleet.

Kamtschatka, South. [293]

1.   dischak.
2.   kascha.
3.   tschook.
4.   tschaaka.
5.   kumnaka.
6.   ky'lkoka.
7.   itatyk   = 2 + 5.
8.   tschookotuk   = 3 + 5.
9.   tschuaktuk   = 4 + 5.
10.   kumechtuk   = 5 + 5.

Aleuts [294]

1.   ataqan.
2.   aljak.
3.   qankun.
4.   sitsin.
5.   tsan   = my hand.
6.   atun   = 1 + 5.
7.   ulun   = 2 + 5.
8.   qamtsin   = 3 + 5.
9.   sitsin   = 4 + 5.
10.   hatsiq.

Tchiglit, Mackenzie R. [295]

1.   ataotçirkr.
2.   aypak, or malloerok.

3.   illaak, or piñatcut.

4.   tçitamat.

5.   tallemat.

6.   arveneloerit.

7.   arveneloerit-aypak   = 5 + 2.

8.   arveneloerit-illaak   = 5 + 3.

9.   arveneloerit-tçitamat   = 5 + 4.

10.   krolit.

Sahaptin (Nez Perces). [296]

1.   naks.

2.   lapit.

3.   mitat.

4.   pi-lapt   = 2 × 2.

5.   pachat.

6.   oi-laks   = [5] + 1.

7.   oi-napt   = [5] + 2.

8.   oi-matat   = [5] + 3.

9.   koits.

10.   putimpt.

Greenland. [297]

1.   atauseq.

2.   machdluq.

3.   pinasut.

4.   sisamat

5.   tadlimat.

6.   achfineq-atauseq   = other hand 1.

7.   achfineq-machdluq   = other hand 2.

8.   achfineq-pinasut   = other hand 3.

9.   achfineq-sisamat   = other hand 4.

10.   qulit.

11.   achqaneq-atauseq   = first foot 1.

12.	achqaneq-machdluq	= first foot 2.
13.	achqaneq-pinasut	= first foot 3.
14.	achqaneq-sisamat	= first foot 4.
15.	achfechsaneq?	
16.	achfechsaneq-atauseq	= other foot 1.
17.	achfechsaneq-machdlup	= other foot 2.
18.	achfechsaneq-pinasut	= other foot 3.
19.	achfechsaneq-sisamat	= other foot 4.
20.	inuk navdlucho	= a man ended.

Up to this point the Greenlander's scale is almost purely quinary. Like those of which mention was made at the beginning of this chapter, it persists in progressing by fives until it reaches 20, when it announces a new base, which shows that the system will from now on be vigesimal. This scale is one of the most interesting of which we have any record, and will be noticed again in the next chapter. In many respects it is like the scale of the Point Barrow Eskimo, which was given early in Chapter III. The Eskimo languages are characteristically quinary-vigesimal in their number systems, but few of them present such perfect examples of that method of counting as do the two just mentioned.

Chippeway. [298]

1.	bejig.	
2.	nij.	
3.	nisswi.	
4.	niwin.	
5.	nanun.	
6.	ningotwasswi	= 1 again?
7.	nijwasswi	= 2 again?
8.	nishwasswi	= 3 again?
9.	jangasswi	= 4 again?
10.	midasswi	= 5 again.

Massachusetts. [299]

1.   nequt.
2.   neese.
3.   nish.
4.   yaw.
5.   napanna   = on one side, *i.e.* 1 hand.
6.   nequttatash   = 1 added.
7.   nesausuk   = 2 again?
8.   shawosuk   = 3 again?
9.   pashoogun   = it comes near, *i.e.* to 10.
10.   puik.

Ojibwa of Chegoimegon. [300]

1.   bashik.
2.   neensh.
3.   niswe.
4.   newin.
5.   nanun.
6.   ningodwaswe   = 1 again?
7.   nishwaswe   = 2 again?
8.   shouswe   = 3 again?
9.   shangaswe   = 4 again?
10.   medaswe   = 5 again?

Ottawa.

1.   ningotchau.
2.   ninjwa.
3.   niswa.
4.   niwin.
5.   nanau.
6.   ningotwaswi   = 1 again?
7.   ninjwaswi   = 2 again?
8.   nichwaswi   = 3 again?
9.   shang.

10.   kwetch.

Delaware.

1.   n'gutti.
2.   niskha.
3.   nakha.
4.   newa.
5.   nalan [akin to palenach, hand].
6.   guttash   = 1 on the other side.
7.   nishash   = 2 on the other side.
8.   khaash   = 3 on the other side.
9.   peshgonk   = coming near.
10.   tellen   = no more.

Shawnoe.

1.   negote.
2.   neshwa.
3.   nithuie.
4.   newe.
5.   nialinwe   = gone.
6.   negotewathwe   = 1 further.
7.   neshwathwe   = 2 further.
8.   sashekswa   = 3 further?
9.   chakatswe [akin to chagisse, "used up"].
10.   metathwe   = no further.

Micmac. [301]

1.   naiookt.
2.   tahboo.
3.   seest.
4.   naioo.
5.   nahn.
6.   usoo-cum.

7.	eloo-igunuk.
8.	oo-gumoolchin.
9.	pescoonaduk.
10.	mtlin.

One peculiarity of the Micmac numerals is most noteworthy. The numerals are real verbs, instead of adjectives, or, as is sometimes the case, nouns. They are conjugated through all the variations of mood, tense, person, and number. The forms given above are not those that would be used in counting, but are for specific use, being varied according to the thought it was intended to express. For example, *naiooktaich* = there is 1, is present tense; *naiooktaichcus*, there was 1, is imperfect; and *encoodaichdedou*, there will be 1, is future. The variation in person is shown by the following inflection:

Present Tense.

1st pers.	tahboosee-ek	= there are 2 of us.
2d pers.	tahboosee-yok	= there are 2 of you.
3d pers.	tahboo-sijik	= there are 2 of them.

Imperfect Tense.

1st pers.	tahboosee-egup	= there were 2 of us.
2d pers.	tahboosee-yogup	= there were 2 of you.
3d pers.	tahboosee-sibunik	= there were 2 of them.

Future Tense.

| 3d pers. | tahboosee-dak | = there will be 2 of them, etc. |

The negative form is also comprehended in the list of possible variations. Thus, *tahboo-seekw*, there are not 2 of them; *mah tahboo-seekw*, there will not be 2 of them; and so on, through all the changes which the conjugation of the verb permits.

Old Algonquin.

1.	peygik.
2.	ninsh.
3.	nisswey.
4.	neyoo.

5.   nahran   = gone.
6.   ningootwassoo   = 1 on the other side.
7.   ninshwassoo   = 2 on the other side.
8.   nisswasso   = 3 on the other side.
9.   shangassoo [akin to chagisse, "used up"].
10. mitassoo   = no further.

Omaha.

1.   meeachchee.
2.   nomba.
3.   rabeenee.
4.   tooba.
5.   satta   = hand, *i.e.* all the fingers turned down.
6.   shappai   = 1 more.
7.   painumba   = fingers 2.
8.   pairabeenee   = fingers 3.
9.   shonka   = only 1 finger (remains).
10. kraibaira   = unbent. [302]

Choctaw.

1.   achofee.
2.   tuklo.
3.   tuchina.
4.   ushta.
5.   tahlape   = the first hand ends.
6.   hanali.
7.   untuklo   = again 2.
8.   untuchina   = again 3.
9.   chokali   = soon the end; *i.e.* next the last.
10. pokoli.

Caddoe.

1.   kouanigh.

2.   behit.
3.   daho.
4.   hehweh.
5.   dihsehkon.
6.   dunkeh.
7.   bisekah   = 5 + 2.
8.   dousehka   = 5 + 3.
9.   hehwehsehka   = 4 + hand.
10.   behnehaugh.

Chippeway.

1.   payshik.
2.   neesh.
3.   neeswoy.
4.   neon.
5.   naman   = gone.
6.   nequtwosswoy   = 1 on the other side.
7.   neeshswosswoy   = 2 on the other side.
8.   swoswoy   = 3 on the other side?
9.   shangosswoy [akin to chagissi, "used up"].
10.   metosswoy   = no further.

Adaize.

1.   nancas.
2.   nass.
3.   colle.
4.   tacache.
5.   seppacan.
6.   pacanancus   = 5 + 1.
7.   pacaness   = 5 + 2.
8.   pacalcon   = 5 + 3.
9.   sickinish   = hands minus?

10.   neusne.

Pawnee.

1.   askoo.
2.   peetkoo.
3.   touweet.
4.   shkeetiksh.
5.   sheeooksh   = hands half.
6.   sheekshabish   = 5 + 1.
7.   peetkoosheeshabish   = 2 + 5.
8.   touweetshabish   = 3 + 5.
9.   looksheereewa   = 10 − 1.
10.   looksheeree   = 2d 5?

Minsi.

1.   gutti.
2.   niskha.
3.   nakba.
4.   newa.
5.   nulan   = gone?
6.   guttash   = 1 added.
7.   nishoash   = 2 added.
8.   khaash   = 3 added.
9.   noweli.
10.   wimbat.

Konlischen.

1.   tlek.
2.   tech.
3.   nezk.
4.   taakun.
5.   kejetschin.
6.   klet uschu   = 5 + 1.

7.	tachate uschu	= 5 + 2.
8.	nesket uschu	= 5 + 3.
9.	kuschok	= 10 − 1?
10.	tschinkat.	

Tlingit. [303]

1.	tlek.	
2.	deq.	
3.	natsk.	
4.	dak'on	= 2d 2.
5.	kedjin	= hand.
6.	tle durcu	= other 1.
7.	daqa durcu	= other 2.
8.	natska durcu	= other 3.
9.	gocuk.	
10.	djinkat	= both hands.

Rapid, or Fall, Indians.

1.	karci.	
2.	neece.	
3.	narce.	
4.	nean.	
5.	yautune.	
6.	neteartuce	= 1 over?
7.	nesartuce	= 2 over?
8.	narswartuce	= 3 over?
9.	anharbetwartuce	= 4 over?
10.	mettartuce	= no further?

Heiltsuk. [304]

1.	men.	
2.	matl.	
3.	yutq.	

4.	mu.	
5.	sky'a.	
6.	katla.	
7.	matlaaus	= other 2?
8.	yutquaus	= other 3?
9.	mamene	= 10 − 1.
10.	aiky'as.	

Nootka. [305]

1.	nup.	
2.	atla.	
3.	katstsa.	
4.	mo.	
5.	sutca.	
6.	nopo	= other 1?
7.	atlpo	= other 2?
8.	atlakutl	= 10 − 2.
9.	ts'owakutl	= 10 − 1.
10.	haiu.	

Tsimshian. [306]

1.	gyak.	
2.	tepqat.	
3.	guant.	
4.	tqalpq.	
5.	kctonc (from *anon*, hand).	
6.	kalt	= 2d 1.
7.	t'epqalt	= 2d 2.
8.	guandalt	= 2d 3?
9.	kctemac.	
10.	gy'ap.	

Bilqula. [306]

1.	(s)maotl.	
2.	tlnos.	
3.	asmost.	
4.	mos.	
5.	tsech.	
6.	tqotl	= 2d 1?
7.	nustlnos	= 2d 2?
8.	k'etlnos	= 2 × 4.
9.	k'esman.	
10.	tskchlakcht.	

Molele. [307]

1.	mangu.	
2.	lapku.	
3.	mutka.	
4.	pipa.	
5.	pika.	
6.	napitka	= 1 + 5.
7.	lapitka	= 2 + 5.
8.	mutpitka	= 3 + 5.
9.	laginstshiatkus.	
10.	nawitspu.	

Waiilatpu. [308]

1.	na.	
2.	leplin.	
3.	matnin.	
4.	piping.	
5.	tawit.	
6.	noina	= [5] + 1.
7.	noilip	= [5] + 2.
8.	noimat	= [5] + 3.
9.	tanauiaishimshim.	

10.	ningitelp.

Lutuami. [307]

1.	natshik.

2.	lapit.

3.	ntani.

4.	wonip.

5.	tonapni.

6.	nakskishuptane	= 1 + 5.

7.	tapkishuptane	= 2 + 5.

8.	ndanekishuptane	= 3 + 5.

9.	natskaiakish	= 10 − 1.

10.	taunip.

Saste (Shasta). [309]

1.	tshiamu.

2.	hoka.

3.	hatski.

4.	irahaia.

5.	etsha.

6.	tahaia.

7.	hokaikinis	= 2 + 5.

8.	hatsikikiri	= 3 + 5.

9.	kirihariki-ikiriu.

10.	etsehewi.

Cahuillo. [310]

1.	supli.

2.	mewi.

3.	mepai.

4.	mewittsu.

5.	nomekadnun.

6.	kadnun-supli	= 5-1.

7.	kan-munwi		= 5-2.
8.	kan-munpa		= 5-3.
9.	kan-munwitsu		= 5-4.
10.	nomatsumi.		

Timukua. [311]

1.	yaha.	
2.	yutsa.	
3.	hapu.	
4.	tseketa.	
5.	marua.	
6.	mareka	= 5 + 1
7.	pikitsa	= 5 + 2
8.	pikinahu	= 5 + 3
9.	peke-tsaketa	= 5 + 4
10.	tuma.	

Otomi [312]

1.	nara.	
2.	yocho.	
3.	chiu.	
4.	gocho.	
5.	kuto.	
6.	rato	= 1 + 5.
7.	yoto	= 2 + 5.
8.	chiato	= 3 + 5.
9.	guto	= 4 + 5.
10.	reta.	

Tarasco. [313]

1.	ma.
2.	dziman.
3.	tanimo.

4.   tamu.

5.   yumu.

6.   kuimu.

7.   yun-dziman   = [5] + 2.

8.   yun-tanimo   = [5] + 3.

9.   yun-tamu   = [5] + 4.

10.   temben.

Matlaltzincan. [314]

1.   indawi.

2.   inawi.

3.   inyuhu.

4.   inkunowi.

5.   inkutaa.

6.   inda-towi   = 1 + 5.

7.   ine-towi   = 2 + 5.

8.   ine-ukunowi   = 2-4.

9.   imuratadahata   = 10 − 1?

10.   inda-hata.

Cora. [315]

1.   ceaut.

2.   huapoa.

3.   huaeica.

4.   moacua.

5.   anxuvi.

6.   a-cevi   = [5] + 1.

7.   a-huapoa   = [5] + 2.

8.   a-huaeica   = [5] + 3.

9.   a-moacua   = [5] + 4.

10.   tamoamata (akin to moamati, "hand").

Aymara. [316]

1.	maya.	
2.	paya.	
3.	kimsa.	
4.	pusi.	
5.	piska.	
6.	tsokta.	
7.	pa-kalko	= 2 + 5.
8.	kimsa-kalko	= 3 + 5.
9.	pusi-kalko	= 4 + 5.
10.	tunka.	

Caribs of Essequibo, Guiana. [317]

1.	oween.	
2.	oko.	
3.	oroowa.	
4.	oko-baimema.	
5.	wineetanee	= 1 hand.
6.	owee-puimapo	= 1 again?
7.	oko-puimapo	= 2 again?
8.	oroowa-puimapo	= 3 again?
9.	oko-baimema-puimapo	= 4 again?
10.	oween-abatoro.	

Carib. [318] (Roucouyenne?)

1.	aban, amoin.	
2.	biama.	
3.	eleoua.	
4.	biam-bouri	= 2 again?
5.	ouacabo-apourcou-aban-tibateli.	
6.	aban laoyagone-ouacabo-apourcou.	
7.	biama laoyagone-ouacabo-apourcou.	
8.	eleoua laoyagone-ouacabo-apourcou.	
9.	− −	

10.   chon noucabo.

It is unfortunate that the meanings of these remarkable numerals cannot be given. The counting is evidently quinary, but the terms used must have been purely descriptive expressions, having their origin undoubtedly in certain gestures or finger motions. The numerals obtained from this region, and from the tribes to the south and east of the Carib country, are especially rich in digital terms, and an analysis of the above numerals would probably show clearly the mental steps through which this people passed in constructing the rude scale which served for the expression of their ideas of number.

Kiriri. [319]

1.	biche.	
2.	watsani.	
3.	watsani dikie.	
4.	sumara oroba.	
5.	mi biche misa	= 1 hand.
6.	mirepri bu-biche misa sai.	
7.	mirepri watsani misa sai.	
8.	mirepri watsandikie misa sai.	
9.	mirepri sumara oraba sai.	
10.	mikriba misa sai	= both hands.

Cayubaba [320]

1.	pebi.	
2.	mbeta.	
3.	kimisa.	
4.	pusi.	
5.	pisika.	
6.	sukuta.	
7.	pa-kaluku	= 2 again?
8.	kimisa-kaluku	= 3 again?
9.	pusu-kaluku	= 4 again?

10.   tunka.

Sapibocona [320]

1.   karata.

2.   mitia.

3.   kurapa.

4.   tsada.

5.   maidara (from *arue,* hand).

6.   karata-rirobo   = 1 hand with.

7.   mitia-rirobo   = 2 hand with.

8.   kurapa-rirobo   = 3 hand with.

9.   tsada-rirobo   = 4 hand with.

10.   bururutse   = hand hand.

Ticuna. [321]

1.   hueih.

2.   tarepueh.

3.   tomepueh.

4.   aguemoujih

5.   hueamepueh.

6.   naïmehueapueh   = 5 + 1.

7.   naïmehueatareh   = 5 + 2.

8.   naïmehueatameapueh   = 5 + 3.

9.   gomeapueh   = 10 − 1.

10.   gomeh.

Yanua. [322]

1.   tckini.

2.   nanojui.

3.   munua.

4.   naïrojuino   = 2d 2.

5.   tenaja.

6.   teki-natea   = 1 again?

7.	nanojui-natea	= 2 again?
8.	munua-natea	= 3 again?
9.	naïrojuino-natea	= 4 again?
10.	huijejuino	= 2 × 5?

The foregoing examples will show with considerable fulness the wide dispersion of the quinary scale. Every part of the world contributes its share except Europe, where the only exceptions to the universal use of the decimal system are the half-dozen languages, which still linger on its confines, whose number base is the vigesimal. Not only is there no living European tongue possessing a quinary number system, but no trace of this method of counting is found in any of the numerals of the earlier forms of speech, which have now become obsolete. The only possible exceptions of which I can think are the Greek πεμπάζειν, to count by fives, and a few kindred words which certainly do hint at a remote antiquity in which the ancestors of the Greeks counted on their fingers, and so grouped their units into fives. The Roman notation, the familiar I., II., III., IV. (originally IIII.), V., VI., etc., with equal certainty suggests quinary counting, but the Latin language contains no vestige of anything of the kind, and the whole range of Latin literature is silent on this point, though it contains numerous references to finger counting. It is quite within the bounds of possibility that the prehistoric nations of Europe possessed and used a quinary numeration. But of these races the modern world knows nothing save the few scanty facts that can be gathered from the stone implements which have now and then been brought to light. Their languages have perished as utterly as have the races themselves, and speculation concerning them is useless. Whatever their form of numeration may have been, it has left no perceptible trace on the languages by which they were succeeded. Even the languages of northern and central Europe which were contemporary with the Greek and Latin of classical times have, with the exception of the Celtic tongues of the extreme North-west, left behind them but meagre traces for the modern student to work on. We presume that the ancient Gauls and Goths, Huns and Scythians, and other barbarian tribes had the same method of numeration that their descendants now have; and it is a matter of certainty that the decimal scale was, at that time, not used with the universality which now obtains; but wherever the decimal was not used, the universal method was vigesimal; and that the quinary ever had anything of a foothold in Europe is only to be guessed from its presence to-day in almost all of the other corners of the world.

From the fact that the quinary is that one of the three natural scales with the smallest base, it has been conjectured that all tribes possess, at some time in their history, a quinary numeration, which at a later period merges into either the decimal or the vigesimal, and thus disappears or forms with one of the latter a mixed system. [323] In support of this theory it is urged that extensive regions which now show nothing but decimal counting were, beyond all reasonable doubt, quinary. It is well known, for example, that the decimal system of the Malays has spread over almost the entire Polynesian region, displacing whatever native scales it encountered. The same phenomenon has been observed in Africa, where the Arab traders have disseminated their own numeral system very widely, the native tribes adopting it or modifying their own scales in such a manner that the Arab influence is detected without difficulty.

In view of these facts, and of the extreme readiness with which a tribe would through its finger counting fall into the use of the quinary method, it does not at first seem improbable that the quinary was *the* original system. But an extended study of the methods of counting in vogue among the uncivilized races of all parts of the world has shown that this theory is entirely untenable. The decimal scale is no less simple in its structure than the quinary; and the savage, as he extends the limit of his scale from 5 to 6, may call his new number 5-1, or, with equal probability, give it an entirely new name, independent in all respects of any that have preceded it. With the use of this new name there may be associated the conception of "5 and 1 more"; but in such multitudes of instances the words employed show no trace of any such meaning, that it is impossible for any one to draw, with any degree of safety, the inference that the signification was originally there, but that the changes of time had wrought changes in verbal form so great as to bury it past the power of recovery. A full discussion of this question need not be entered upon here. But it will be of interest to notice two or three numeral scales in which the quinary influence is so faint as to be hardly discernible. They are found in considerable numbers among the North American Indian languages, as may be seen by consulting the vocabularies that have been prepared and published during the last half century. [324] From these I have selected the following, which are sufficient to illustrate the point in question:

Quappa.

1.	milchtih.
2.	nonnepah.
3.	dahghenih.
4.	tuah.
5.	sattou.
6.	schappeh.
7.	pennapah.
8.	pehdaghenih.
9.	schunkkah.
10.	gedeh bonah.

Terraba. [325]

1.	krara.
2.	krowü.
3.	krom miah.
4.	krob king.
5.	krasch kingde.
6.	terdeh.
7.	kogodeh.
8.	kwongdeh.
9.	schkawdeh.
10.	dwowdeh.

Mohican

1.	ngwitloh.
2.	neesoh.
3.	noghhoh.

4.   nauwoh.

5.   nunon.

6.   ngwittus.

7.   tupouwus.

8.   ghusooh.

9.   nauneeweh.

10.   mtannit.

In the Quappa scale 7 and 8 appear to be derived from 2 and 3, while 6 and 9 show no visible trace of kinship with 1 and 4. In Mohican, on the other hand, 6 and 9 seem to be derived from 1 and 4, while 7 and 8 have little or no claim to relationship with 2 and 3. In some scales a single word only is found in the second quinate to indicate that 5 was originally the base on which the system rested. It is hardly to be doubted, even, that change might affect each and every one of the numerals from 5 to 10 or 6 to 9, so that a dependence which might once have been easily detected is now unrecognizable.

But if this is so, the natural and inevitable question follows — might not this have been the history of all numeral scales now purely decimal? May not the changes of time have altered the compounds which were once a clear indication of quinary counting, until no trace remains by which they can be followed back to their true origin? Perhaps so. It is not in the least degree probable, but its possibility may, of course, be admitted. But even then the universality of quinary counting for primitive peoples is by no means established. In Chapter II, examples were given of races which had no number base. Later on it was observed that in Australia and South America many tribes used 2 as their number base; in some cases counting on past 5 without showing any tendency to use that as a new unit. Again, through the habit of counting upon the finger joints, instead of the fingers themselves, the use of 3 as a base is brought into prominence, and 6 and 9 become 2 threes and 3 threes, respectively, instead of 5 + 1 and 5 + 4. The same may be noticed of 4. Counting by means of his fingers, without including the thumbs, the savage begins by dividing into fours instead of fives. Traces of this form of counting are somewhat numerous, especially among the North American aboriginal tribes. Hence the quinary form of counting, however widespread its use may be shown to be, can in no way be claimed as the universal method of any stage of development in the history of mankind.

In the vast majority of cases, the passage from the base to the next succeeding number in any scale, is clearly defined. But among races whose intelligence is of a low order, or—if it be permissible to express it in this way—among races whose number sense is feeble, progression from one number to the next is not always in accordance with any well-defined law. After one or two distinct numerals the count may, as in the case of the Veddas and the Andamans, proceed by finger pantomime and by the repetition of the same word. Occasionally the same word is used for two successive numbers, some gesture undoubtedly serving to distinguish the one from the other in the savage's mind. Examples of this are not infrequent among the forest tribes of South America. In the Tariana dialect 9 and 10 are expressed by the same word, *paihipawalianuda;* in Cobeu, 8 and 9 by *pepelicoloblicouilini;* in Barre, 4, 5, and 9 by *ualibucubi.* [326] In other languages the change from one numeral to the next is so slight that one instinctively concludes that the savage is forming in his own mind another, to him new, numeral immediately from the last. In such cases the entire number system is scanty, and the creeping hesitancy with which progress is made is visible in the forms which the numerals are made to take. A single illustration or two of this must suffice; but the ones chosen are not isolated cases. The scale of the Macunis, [327] one of the numerous tribes of Brazil, is

1. pocchaenang.
2. haihg.
3. haigunhgnill.
4. haihgtschating.
5. haihgtschihating   = another 4?
6. hathig-stchihathing   = 2-4?
7. hathink-tschihathing   = 2-5?
8. hathink-tschihating   = 2 × 4?

The complete absence of—one is tempted to say—any rhyme or reason from this scale is more than enough to refute any argument which might tend to show that the quinary, or any other scale, was ever the sole number scale of primitive man. Irregular as this is, the system of the Montagnais fully matches it, as the subjoined numerals show: [328]

1.	inl'are.	
2.	nak'e.	
3.	t'are.	
4.	dinri.	
5.	se-sunlare.	
6.	elkke-t'are	$= 2 \times 3.$
7.	t'a-ye-oyertan	$= 10 - 3,$
	or inl'as dinri	$= 4 + 3?$
8.	elkke-dinri	$= 2 \times 4.$
9.	inl'a-ye-oyertan	$= 10 - 1.$
10.	onernan.	

Chapter VII.

The Vigesimal System.

In its ordinary development the quinary system is almost sure to merge into either the decimal or the vigesimal system, and to form, with one or the other or both of these, a mixed system of counting. In Africa, Oceanica, and parts of North America, the union is almost always with the decimal scale; while in other parts of the world the quinary and the vigesimal systems have shown a decided affinity for each other. It is not to be understood that any geographical law of distribution has ever been observed which governs this, but merely that certain families of races have shown a preference for the one or the other method of counting. These families, disseminating their characteristics through their various branches, have produced certain groups of races which exhibit a well-marked tendency, here toward the decimal, and there toward the vigesimal form of numeration. As far as can be ascertained, the choice of the one or the other scale is determined by no external circumstances, but depends solely on the mental characteristics of the tribes themselves. Environment does not exert any appreciable influence either. Both decimal and vigesimal numeration are found indifferently in warm and in cold countries; in fruitful and in barren lands; in maritime and in inland regions; and among highly civilized or deeply degraded peoples.

Whether or not the principal number base of any tribe is to be 20 seems to depend entirely upon a single consideration; are the fingers alone used as an aid to counting, or are both fingers and toes used? If only the fingers are employed, the resulting scale must become decimal if sufficiently extended. If use is made of the toes in addition to the fingers, the outcome must inevitably be a vigesimal system. Subordinate to either one of these the quinary may and often does appear. It is never the principal base in any extended system.

To the statement just made respecting the origin of vigesimal counting, exception may, of course, be taken. In the case of numeral scales like the Welsh, the Nahuatl, and many others where the exact meanings of the numerals cannot be ascertained, no proof exists that the ancestors of these peoples ever used either finger or toe counting; and the sweeping statement that any vigesimal scale is the outgrowth of the use of these natural counters is not susceptible of proof. But so many examples are met with in which the origin is clearly of this nature, that no hesitation is felt in putting the above forward as a general explanation for the existence of this kind of counting. Any other origin is difficult to reconcile with observed facts, and still more difficult to reconcile with any rational theory of number system development. Dismissing from consideration the quinary scale, let us briefly examine once more the natural process of evolution through which the decimal and the vigesimal scales come into being. After the completion of one count of the fingers the savage announces his result in some form which definitely states to his mind the fact that the end of a well-marked series has been reached. Beginning again, he now repeats his count of 10, either on his own fingers or on the fingers of another. With the completion of the second 10 the result is announced, not in a new unit, but by means of a duplication of the term already used. It is scarcely credible that the unit unconsciously adopted at the termination of the first count should now be dropped, and a new one substituted in its place. When the method here described is employed, 20 is not a natural unit to which higher numbers may be referred. It is wholly artificial; and it would be most surprising if it were adopted. But if the count of the second 10 is made on the toes in place of the fingers, the element of repetition which entered into the previous method is now wanting. Instead of referring each new number to the 10 already completed, the savage is still feeling his way along, designating his new terms by such phrases as "1 on the foot," "2 on the other foot," etc. And now, when 20 is reached, a single series is finished instead of a double series as before; and the result is expressed in one of the many methods already noticed—"one man," "hands and feet," "the feet finished," "all the fingers of hands and feet," or some equivalent formula. Ten is no longer the natural base. The number from which the new start is made is 20, and the resulting scale is inevitably vigesimal. If pebbles or sticks are used instead of fingers, the system will probably be decimal. But back of the stick and pebble counting the 10 natural counters always exist, and to them we must always look for the origin of this scale.

In any collection of the principal vigesimal number systems of the world, one would naturally begin with those possessed by the Celtic races of Europe. These races, the earliest European peoples of whom we have any exact knowledge, show a preference for counting by twenties, which is almost as decided as that manifested by Teutonic races for counting by tens. It has been conjectured by some writers that the explanation for this was to be found in the ancient commercial intercourse which existed between the Britons and the Carthaginians and Phœnicians, whose number systems showed traces of a vigesimal tendency. Considering the fact that the use of vigesimal counting was universal among Celtic races, this explanation is quite gratuitous. The reason why the Celts used this method is entirely unknown, and need not concern investigators in the least. But the fact that they did use it is important, and commands attention. The five Celtic languages, Breton, Irish, Welsh, Manx, and Gaelic, contain the following well-defined vigesimal scales. Only the principal or characteristic numerals are given, those being sufficient to enable the reader to follow intelligently the growth of the systems. Each contains the decimal element also, and is, therefore, to be regarded as a mixed decimal-vigesimal system.

Irish. [329]

10.	deic.	
20.	fice.	
30.	triocad	= 3-10
40.	da ficid	= 2-20.
50.	caogad	= 5-10.
60.	tri ficid	= 3-20.
70.	reactmoga	= 7-10.
80.	ceitqe ficid	= 4-20.
90.	nocad	= 9-10.
100.	cead.	
1000.	mile.	

Gaelic. [330]

10.	deich.	
20.	fichead.	
30.	deich ar fichead	= 10 + 20.
40.	da fhichead	= 2-20.
50.	da fhichead is deich	= 40 + 10.
60.	tri fichead	= 3-20.
70.	tri fichead is deich	= 60 + 10.
80.	ceithir fichead	= 4-20.
90.	ceithir fichead is deich	= 80 + 10.
100.	ceud.	
1000.	mile.	

Welsh. [331]

10.	deg.	
20.	ugain.	
30.	deg ar hugain	= 10 + 20.
40.	deugain	= 2-20.
50.	deg a deugain	= 10 + 40.
60.	trigain	= 3-20.
70.	deg a thrigain	= 10 + 60.
80.	pedwar ugain	= 4-20.
90.	deg a pedwar ugain	= 80 + 10.
100.	cant.	

Manx. [332]

10.	jeih.	
20.	feed.	
30.	yn jeih as feed	= 10 + 20.
40.	daeed	= 2-20.
50.	jeih as daeed	= 10 + 40.
60.	three-feed	= 3-20.

70.	three-feed as jeih	= 60 + 10.
80.	kiare-feed	= 4-20.
100.	keead.	
1000.	thousane, or jeih cheead.	

Breton. [333]

10.	dec.	
20.	ueguend.	
30.	tregond	= 3-10.
40.	deu ueguend	= 2-20.
50.	hanter hand	= half hundred.
60.	tri ueguend	= 3-20.
70.	dec ha tri ueguend	= 10 + 60.
80.	piar ueguend	= 4-20.
90.	dec ha piar ueguend	= 10 + 80.
100.	cand.	
120.	hueh ueguend	= 6-20.
140.	seih ueguend	= 7-20.
160.	eih ueguend	= 8-20.
180.	nau ueguend	= 9-20.
200.	deu gand	= 2-100.
240.	deuzec ueguend	= 12-20.
280.	piarzec ueguend	= 14-20.
300.	tri hand, or pembzec ueguend.	
400.	piar hand	= 4-100.
1000.	mil.	

These lists show that the native development of the Celtic number systems, originally showing a strong preference for the vigesimal method of progression, has been greatly modified by intercourse with Teutonic and Latin races. The higher numerals in all these languages, and in Irish many of the lower also, are seen at a glance to be decimal. Among the scales here given the Breton, the legitimate descendant of the ancient Gallic, is especially interesting; but here, just as in the other Celtic tongues, when we reach 1000, the familiar Latin term for that number appears in the various corruptions of *mille*, 1000, which was carried into the Celtic countries by missionary and military influences.

In connection with the Celtic language, mention must be made of the persistent vigesimal element which has held its place in French. The ancient Gauls, while adopting the language of their conquerors, so far modified the decimal system of Latin as to replace the natural *septante*, 70, *octante*, 80, *nonante*, 90, by *soixante-dix*, 60-10, *quatre-vingt*, 4-20, and *quatrevingt-dix*, 4-20-10. From 61 to 99 the French method of counting is wholly vigesimal, except for the presence of the one word *soixante*. In old French this element was still more pronounced. *Soixante* had not yet appeared; and 60 and 70 were *treis vinz*, 3-20, and *treis vinz et dis*, 3-20 and 10 respectively. Also, 120 was *six vinz*, 6-20, 140 was *sept-vinz*, etc. [334] How far this method ever extended in the French language proper, it is, perhaps, impossible to say; but from the name of an almshouse, *les quinze-vingts*, [335] which formerly existed in Paris, and was designed as a home for 300 blind persons, and from the *pembzek-ueguent*, 15-20, of the Breton, which still survives, we may infer that it was far enough to make it the current system of common life.

Europe yields one other example of vigesimal counting, in the number system of the Basques. Like most of the Celtic scales, the Basque seems to become decimal above 100. It does not appear to be related to any other European system, but to be quite isolated philologically. The higher units, as *mila*, 1000, are probably borrowed, and not native. The tens in the Basque scale are: [336]

10.	hamar.	
20.	hogei.	
30.	hogei eta hamar	= 20 + 10.
40.	berrogei	= 2-20.
50.	berrogei eta hamar	= 2-20 + 10.
60.	hirurogei	= 3-20.
70.	hirurogei eta hamar	= 3-20 + 10.
80.	laurogei	= 4-20.

90.	laurogei eta hamar	= 4-20 + 10.
100.	ehun.	
1000.	*milla.*	

Besides these we find two or three numeral scales in Europe which contain distinct traces of vigesimal counting, though the scales are, as a whole, decidedly decimal. The Danish, one of the essentially Germanic languages, contains the following numerals:

30.	tredive	= 3-10.
40.	fyrretyve	= 4-10.
50.	halvtredsindstyve	= half (of 20) from 3-20.
60.	tresindstyve	= 3-20.
70.	halvfierdsindstyve	= half from 4-20.
80.	fiirsindstyve	= 4-20.
90.	halvfemsindstyve	= half from 5-20.
100.	hundrede.	

Germanic number systems are, as a rule, pure decimal systems; and the Danish exception is quite remarkable. We have, to be sure, such expressions in English as *three score*, *four score*, etc., and the Swedish, Icelandic, and other languages of this group have similar terms. Still, these are not pure numerals, but auxiliary words rather, which belong to the same category as *pair*, *dozen*, *dizaine*, etc., while the Danish words just given are the ordinary numerals which form a part of the every-day vocabulary of that language. The method by which this scale expresses 50, 70, and 90 is especially noticeable. It will be met with again, and further examples of its occurrence given.

In Albania there exists one single fragment of vigesimal numeration, which is probably an accidental compound rather than the remnant of a former vigesimal number system. With this single exception the Albanian scale is of regular decimal formation. A few of the numerals are given for the sake of comparison: [337]

| 30. | tridgiete | = 3-10. |
| 40. | dizet | = 2-20. |

| 50. | pesedgiete | = 5-10. |
| 60. | giastedgiete | = 6-10, etc. |

Among the almost countless dialects of Africa we find a comparatively small number of vigesimal number systems. The powers of the negro tribes are not strongly developed in counting, and wherever their numeral scales have been taken down by explorers they have almost always been found to be decimal or quinary-decimal. The small number I have been able to collect are here given. They are somewhat fragmentary, but are as complete as it was possible to make them.

Affadeh. [338]

10.	dekang.	
20.	degumm.	
30.	piaske.	
40.	tikkumgassih	= 20 × 2.
50.	tikkumgassigokang	= 20 × 2 + 10.
60.	tikkumgakro	= 20 × 3.
70.	dungokrogokang	= 20 × 3 + 10.
80.	dukumgade	= 20 × 4.
90.	dukumgadegokang	= 20 × 4 + 10.
100.	miah (borrowed from the Arabs).	

Ibo. [339]

10.	iri.	
20.	ogu.	
30.	ogu n-iri	= 20 + 10,
	or iri ato	= 10 × 3.
40.	ogu abuo	= 20 × 2,
	or iri anno	= 10 × 4.
100.	ogu ise	= 20 × 5.

Vei. [340]

| 10. | tan. | |

20.	mo bande	= a person finished.
30.	mo bande ako tan	= 20 + 10.
40.	mo fera bande	= 2 × 20.
100.	mo soru bande	= 5 persons finished.

Yoruba. [341]

10.	duup.	
20.	ogu.	
30.	ogbo.	
40.	ogo-dzi	= 20 × 2.
60.	ogo-ta	= 20 × 3.
80.	ogo-ri	= 20 × 4.
100.	ogo-ru	= 20 × 5.
120.	ogo-fa	= 20 × 6.
140.	ogo-dze	= 20 × 7.
160.	ogo-dzo	= 20 × 8, etc.

Efik. [342]

10.	duup.	
20.	edip.	
30.	edip-ye-duup	= 20 + 10.
40.	aba	= 20 × 2.
60.	ata	= 20 × 3.
80.	anan	= 20 × 4.
100.	ikie.	

The Yoruba scale, to which reference has already been made, p. 70, again shows its peculiar structure, by continuing its vigesimal formation past 100 with no interruption in its method of numeral building. It will be remembered that none of the European scales showed this persistency, but passed at that point into decimal numeration. This will often be found to be the case; but now and then a scale will come to our notice whose vigesimal structure is continued, without any break, on into the hundreds and sometimes into the thousands.

Bongo. [343]

10.	kih.	
20.	mbaba kotu	$= 20 \times 1.$
40.	mbaba gnorr	$= 20 \times 2.$
100.	mbaba mui	$= 20 \times 5.$

Mende. [344]

10.	pu.	
20.	nu yela gboyongo mai	= a man finished.
30.	nu yela gboyongo mahu pu	$= 20 + 10.$
40.	nu fele gboyongo	= 2 men finished.
100.	nu lolu gboyongo	= 5 men finished.

Nupe. [345]

10.	gu-wo.	
20.	esin.	
30.	gbonwo.	
40.	si-ba	$= 2 \times 20.$
50.	arota.	
60.	sita	$= 3 \times 20.$
70.	adoni.	
80.	sini	$= 4 \times 20.$
90.	sini be-guwo	$= 80 + 10.$
100.	sisun	$= 5 \times 20.$

Logone. [346]

10.	chkan.	
20.	tkam.	
30.	tkam ka chkan	$= 20 + 10.$
40.	tkam ksde	$= 20 \times 2.$
50.	tkam ksde ka chkan	$= 40 + 10.$
60.	tkam gachkir	$= 20 \times 3.$
100.	mia (from Arabic).	

1000.   debu.

Mundo. [347]

10.	nujorquoi.	
20.	tiki bere.	
30.	tiki bire nujorquoi	= 20 + 10.
40.	tiki borsa	= 20 × 2.
50.	tike borsa nujorquoi	= 40 + 10.

Mandingo. [348]

10.	tang.	
20.	mulu.	
30.	mulu nintang	= 20 + 10.
40.	mulu foola	= 20 × 2.
50.	mulu foola nintang	= 40 + 10.
60.	mulu sabba	= 20 × 3.
70.	mulu sabba nintang	= 60 + 10.
80.	mulu nani	= 20 × 4.
90.	mulu nani nintang	= 80 + 10.
100.	kemi.	

This completes the scanty list of African vigesimal number systems that a patient and somewhat extended search has yielded. It is remarkable that the number is no greater. Quinary counting is not uncommon in the "Dark Continent," and there is no apparent reason why vigesimal reckoning should be any less common than quinary. Any one investigating African modes of counting with the material at present accessible, will find himself hampered by the fact that few explorers have collected any except the first ten numerals. This leaves the formation of higher terms entirely unknown, and shows nothing beyond the quinary or non-quinary character of the system. Still, among those which Stanley, Schweinfurth, Salt, and others have collected, by far the greatest number are decimal. As our knowledge of African languages is extended, new examples of the vigesimal method may be brought to light. But our present information leads us to believe that they will be few in number.

In Asia the vigesimal system is to be found with greater frequency than in Europe or Africa, but it is still the exception. As Asiatic languages are much better known than African, it is probable that the future will add but little to our stock of knowledge on this point. New instances of counting by twenties may still be found in northern Siberia, where much ethnological work yet remains to be done, and where a tendency toward this form of numeration has been observed to exist. But the total number of Asiatic vigesimal scales must always remain small — quite insignificant in comparison with those of decimal formation.

In the Caucasus region a group of languages is found, in which all but three or four contain vigesimal systems. These systems are as follows:

Abkhasia. [349]

10.	zpha-ba.	
20.	gphozpha	$= 2 \times 10.$
30.	gphozphei zphaba	$= 20 + 10.$
40.	gphin-gphozpha	$= 2 \times 20.$
60.	chin-gphozpha	$= 3 \times 20.$
80.	phsin-gphozpha	$= 4 \times 20.$
100.	sphki.	

Avari

10.	antsh-go.	
20.	qo-go.	
30.	lebergo.	
40.	khi-qogo	$= 2 \times 20.$
50.	khiqojalda antshgo	$= 40 + 10.$
60.	lab-qogo	$= 3 \times 20.$
70.	labqojalda antshgo	$= 60 + 10.$
80.	un-qogo	$= 4 \times 20.$
100.	nusgo.	

Kuri

10.	tshud.	
20.	chad.	
30.	channi tshud	= 20 + 10.
40.	jachtshur.	
50.	jachtshurni tshud	= 40 + 10.
60.	put chad	= 3 × 20.
70.	putchanni tshud	= 60 + 10.
80.	kud-chad	= 4 × 20.
90.	kudchanni tshud	= 80 + 10.
100.	wis.	

Udi

10.	witsh.	
20.	qa.	
30.	sa-qo-witsh	= 20 + 10.
40.	pha-qo	= 2 × 20.
50.	pha-qo-witsh	= 40 + 10.
60.	chib-qo	= 3 × 20.
70.	chib-qo-witsh	= 60 + 10.
80.	bip-qo	= 4 × 20.
90.	bip-qo-witsh	= 80 + 10.
100.	bats.	
1000.	hazar (Persian).	

Tchetchnia

10.	ith.	
20.	tqa.	
30.	tqe ith	= 20 + 10.
40.	sauz-tqa	= 2 × 20.
50.	sauz-tqe ith	= 40 + 10.
60.	chuz-tqa	= 3 × 20.
70.	chuz-tqe ith	= 60 + 10.
80.	w-iez-tqa	= 4 × 20.

90.	w-iez-tqe ith	= 80 + 10.
100.	b'e.	
1000.	ezir (akin to Persian).	

Thusch

10.	itt.	
20.	tqa.	
30.	tqa-itt	= 20 + 10.
40.	sauz-tq	= 2 × 20.
50.	sauz-tqa-itt	= 40 + 10.
60.	chouz-tq	= 3 × 20.
70.	chouz-tqa-itt	= 60 + 10.
80.	dhewuz-tq	= 4 × 20.
90.	dhewuz-tqa-itt	= 80 + 10.
100.	phchauz-tq	= 5 × 20.
200.	itsha-tq	= 10 × 20.
300.	phehiitsha-tq	= 15 × 20.
1000.	satsh tqauz-tqa itshatqa	= 2 × 20 × 20 + 200.

Georgia

10.	athi.	
20.	otsi.	
30.	ots da athi	= 20 + 10.
40.	or-m-otsi	= 2 × 20.
50.	ormots da athi	= 40 + 10.
60.	sam-otsi	= 3 × 20.
70.	samots da athi	= 60 + 10.
80.	othch-m-otsi	= 4 × 20.
90.	othmots da athi	= 80 + 10.
100.	asi.	
1000.	ath-asi	= 10 × 100.

Lazi

10.	wit.	
20.	öts.	
30.	öts do wit	= 20 × 10.
40.	dzur en öts	= 2 × 20.
50.	dzur en öts do wit	= 40 + 10.
60.	dzum en öts	= 3 × 20.
70.	dzum en öts do wit	= 60 + 10.
80.	otch-an-öts	= 4 × 20.
100.	os.	
1000.	silia (akin to Greek).	

Chunsag. [350]

10.	ants-go.	
20.	chogo.	
30.	chogela antsgo	= 20 + 10.
40.	kichogo	= 2 × 20.
50.	kichelda antsgo	= 40 + 10.
60.	taw chago	= 3 × 20.
70.	taw chogelda antsgo	= 60 + 10.
80.	uch' chogo	= 4 × 20.
90.	uch' chogelda antsgo.	
100.	nusgo.	
1000.	asargo (akin to Persian).	

Dido. [351]

10.	zino.	
20.	ku.	
30.	kunozino.	
40.	kaeno ku	= 2 × 20.
50.	kaeno kuno zino	= 40 + 10.
60.	sonno ku	= 3 × 20.
70.	sonno kuno zino	= 60 + 10.
80.	uino ku	= 4 × 20.

90.	uino huno zino	= 80 + 10.
100.	bischon.	
400.	kaeno kuno zino	= 40 × 10.

Akari

10.	entzelgu.	
20.	kobbeggu.	
30.	lowergu.	
40.	kokawu	= 2 × 20.
50.	kikaldanske	= 40 + 10.
60.	secikagu.	
70.	kawalkaldansku	= 3 × 20 + 10.
80.	onkuku	= 4 × 20.
90.	onkordansku	= 4 × 20 + 10.
100.	nosku.	
1000.	askergu (from Persian).	

Circassia

10.	psche.	
20.	to-tsch.	
30.	totsch-era-pschirre	= 20 + 10.
40.	ptl'i-sch	= 4 × 10.
50.	ptl'isch-era-pschirre	= 40 + 10.
60.	chi-tsch	= 6 × 10.
70.	chitsch-era-pschirre	= 60 + 10.
80.	toshitl	= 20 × 4?
90.	toshitl-era-pschirre	= 80 + 10.
100.	scheh.	
1000.	min (Tartar) or schi-psche	= 100 × 10.

The last of these scales is an unusual combination of decimal and vigesimal. In the even tens it is quite regularly decimal, unless 80 is of the structure suggested above. On the other hand, the odd tens are formed in the ordinary vigesimal manner. The reason for this anomaly is not obvious. I know of no other number system that presents the same peculiarity, and cannot give any hypothesis which will satisfactorily account for its presence here. In nearly all the examples given the decimal becomes the leading element in the formation of all units above 100, just as was the case in the Celtic scales already noticed.

Among the northern tribes of Siberia the numeral scales appear to be ruder and less simple than those just examined, and the counting to be more consistently vigesimal than in any scale we have thus far met with. The two following examples are exceedingly interesting, as being among the best illustrations of counting by twenties that are to be found anywhere in the Old World.

Tschukschi. [352]

10.	migitken	= both hands.
20.	chlik-kin	= a whole man.
30.	chlikkin mingitkin parol	= 20 + 10.
40.	nirach chlikkin	= 2 × 20.
100.	milin chlikkin	= 5 × 20.
200.	mingit chlikkin	= 10 × 20, *i.e.* 10 men.
1000.	miligen chlin-chlikkin	= 5 × 200, *i.e.* five (times) 10 men.

Aino. [353]

10.	wambi.	
20.	choz.	
30.	wambi i-doehoz	= 10 from 40.
40.	tochoz	= 2 × 20.
50.	wambi i-richoz	= 10 from 60.
60.	rechoz	= 3 × 20.
70.	wambi [i?] inichoz	= 10 from 80.
80.	inichoz	= 4 × 20.
90.	wambi aschikinichoz	= 10 from 100.
100.	aschikinichoz	= 5 × 20.
110.	wambi juwanochoz	= 10 from 120.

120.	juwano choz	$= 6 \times 20.$
130.	wambi aruwanochoz	$= 10$ from 140.
140.	aruwano choz	$= 7 \times 20.$
150.	wambi tubischano choz	$= 10$ from 160.
160.	tubischano choz	$= 8 \times 20.$
170.	wambi schnebischano choz	$= 10$ from 180.
180.	schnebischano choz	$= 9 \times 20.$
190.	wambi schnewano choz	$= 10$ from 200.
200.	schnewano choz	$= 10 \times 20.$
300.	aschikinichoz i gaschima chnewano choz	$= 5 \times 20 + 10 \times 20.$
400.	toschnewano choz	$= 2 \times (10 \times 20).$
500.	aschikinichoz i gaschima toschnewano choz	$= 100 + 400.$
600.	reschiniwano choz	$= 3 \times 200.$
700.	aschikinichoz i gaschima reschiniwano choz	$= 100 + 600.$
800.	inischiniwano choz	$= 4 \times 200.$
900.	aschikinichoz i gaschima inischiniwano choz	$= 100 + 800.$
1000.	aschikini schinewano choz	$= 5 \times 200.$
2000.	wanu schinewano choz	$= 10 \times (10 \times 20).$

This scale is in one sense wholly vigesimal, and in another way it is not to be regarded as pure, but as mixed. Below 20 it is quinary, and, however far it might be extended, this quinary element would remain, making the scale quinary-vigesimal. But in another sense, also, the Aino system is not pure. In any unmixed vigesimal scale the word for 400 must be a simple word, and that number must be taken as the vigesimal unit corresponding to 100 in the decimal scale. But the Ainos have no simple numeral word for any number above 20, forming all higher numbers by combinations through one or more of the processes of addition, subtraction, and multiplication. The only number above 20 which is used as a unit is 200, which is expressed merely as 10 twenties. Any even number of hundreds, or any number of thousands, is then indicated as being so many times 10 twenties; and the odd hundreds are so many times 10 twenties, plus 5 twenties more. This scale is an excellent example of the cumbersome methods used by uncivilized races in extending their number systems beyond the ordinary needs of daily life.

In Central Asia a single vigesimal scale comes to light in the following fragment of the Leptscha scale, of the Himalaya region: [354]

10.	kati.	
40.	kafali	= 4 × 10,
	or kha nat	= 2 × 20.
50.	kafano	= 5 × 10,
	or kha nat sa kati	= 2 × 20 + 10.
100.	gjo, or kat.	

Further to the south, among the Dravidian races, the vigesimal element is also found. The following will suffice to illustrate the number systems of these dialects, which, as far as the material at hand shows, are different from each other only in minor particulars:

Mundari. [355]

10.	gelea.	
20.	mi hisi.	
30.	mi hisi gelea	= 20 + 10.
40.	bar hisi	= 2 × 20.
60.	api hisi	= 3 × 20.
80.	upun hisi	= 4 × 20.
100.	mone hisi	= 5 × 20.

In the Nicobar Islands of the Indian Ocean a well-developed example of vigesimal numeration is found. The inhabitants of these islands are so low in the scale of civilization that a definite numeral system of any kind is a source of some surprise. Their neighbours, the Andaman Islanders, it will be remembered, have but two numerals at their command; their intelligence does not seem in any way inferior to that of the Nicobar tribes, and one is at a loss to account for the superior development of the number sense in the case of the latter. The intercourse of the coast tribes with traders might furnish an explanation of the difficulty were it not for the fact that the numeration of the inland tribes is quite as well developed as that of the coast tribes; and as the former never come in contact with traders and never engage in barter of any kind except in the most limited way, the conclusion seems inevitable that this is merely one of the phenomena of mental development among savage races for which we have at present no adequate explanation. The principal numerals of the inland and of the coast tribes are: [356]

Inland Tribes

10.	teya.	
20.	heng-inai.	
30.	heng-inai-tain	$= 20 + 5$ (couples).
40.	au-inai	$= 2 \times 20.$
100.	tain-inai	$= 5 \times 20.$
200.	teya-inai	$= 10 \times 20.$
300.	teya-tain-inai	$= (10 + 5) \times 20.$
400.	heng-teo.	

Coast Tribes

10.	sham.	
20.	heang-inai.	
30.	heang-inai-tanai	$= 20 + 5$ (couples).
40.	an-inai	$= 2 \times 20.$
100.	tanai-inai	$= 5 \times 20.$
200.	sham-inai	$= 10 \times 20.$
300.	heang-tanai-inai	$= (10 + 5)\, 20.$
400.	heang-momchiama.	

In no other part of the world is vigesimal counting found so perfectly developed, and, among native races, so generally preferred, as in North and South America. In the eastern portions of North America and in the extreme western portions of South America the decimal or the quinary decimal scale is in general use. But in the northern regions of North America, in western Canada and north-western United States, in Mexico and Central America, and in the northern and western parts of South America, the unit of counting among the great majority of the native races was 20. The ethnological affinities of these races are not yet definitely ascertained; and it is no part of the scope of this work to enter into any discussion of that involved question. But either through contact or affinity, this form of numeration spread in prehistoric times over half or more than half of the western hemisphere. It was the method employed by the rude Eskimos of the north and their equally rude kinsmen of Paraguay and eastern Brazil; by the forest Indians of Oregon and British Columbia, and by their more southern kinsmen, the wild tribes of the Rio Grande and of the Orinoco. And, most striking and interesting of all, it was the method upon which were based the numeral systems of the highly civilized races of Mexico, Yucatan, and New Granada. Some of the systems obtained from the languages of these peoples are perfect, extended examples of vigesimal counting, not to be duplicated in any other quarter of the globe. The ordinary unit was, as would be expected, "one man," and in numerous languages the words for 20 and man are identical. But in other cases the original meaning of that numeral word has been lost; and in others still it has a signification quite remote from that given above. These meanings will be noticed in connection with the scales themselves, which are given, roughly speaking, in their geographical order, beginning with the Eskimo of the far north. The systems of some of the tribes are as follows:

Alaskan Eskimos. [357]

10.	koleet.	
20.	enuenok.	
30.	enuenok kolinik	= 20 + 10.
40.	malho kepe ak	= 2 × 20.
50.	malho-kepe ak-kolmik che pah ak to	= 2 × 20 + 10.
60.	pingi shu-kepe ak	= 3 × 20.
100.	tale ma-kepe ak	= 5 × 20.
400.	enue nok ke pe ak	= 20 × 20.

Tchiglit. [358]

10.	krolit.

20.	kroleti, or innun	= man.
30.	innok krolinik-tchikpalik	= man + 2 hands.
40.	innum mallerok	= 2 men.
50.	adjigaynarmitoat	= as many times 10 as the fingers of the hand.
60.	innumipit	= 3 men.
70.	innunmalloeronik ar-veneloerit	= 7 men?
80.	innun pinatçunik ar-veneloerit	= 8 men?
90.	innun tcitamanik ar-veneloerit	= 9 men?
100.	itchangnerkr.	
1000.	itchangner-park	= great 100.

The meanings for 70, 80, 90, are not given by Father Petitot, but are of such a form that the significations seem to be what are given above. Only a full acquaintance with the Tchiglit language would justify one in giving definite meanings to these words, or in asserting that an error had been made in the numerals. But it is so remarkable and anomalous to find the decimal and vigesimal scales mingled in this manner that one involuntarily suspects either incompleteness of form, or an actual mistake.

Tlingit. [359]

10.	djinkat	= both hands?
20.	tle ka	= 1 man.
30.	natsk djinkat	$= 3 \times 10$.
40.	dak'on djinkat	$= 4 \times 10$.
50.	kedjin djinkat	$= 5 \times 10$.
60.	tle durcu djinkat	$= 6 \times 10$.
70.	daqa durcu djinkat	$= 7 \times 10$.
80.	natska durcu djinkat	$= 8 \times 10$.
90.	gocuk durcu djinkat	$= 9 \times 10$.

100.	kedjin ka	= 5 men, or 5 × 20.
200.	djinkat ka	= 10 × 20.
300.	natsk djinkat ka	= 30 men.
400.	dak'on djinkat ka	= 40 men.

This scale contains a strange commingling of decimal and vigesimal counting. The words for 20, 100, and 200 are clear evidence of vigesimal, while 30 to 90, and the remaining hundreds, are equally unmistakable proof of decimal, numeration. The word *ka*, man, seems to mean either 10 or 20; a most unusual occurrence. The fact that a number system is partly decimal and partly vigesimal is found to be of such frequent occurrence that this point in the Tlingit scale need excite no special wonder. But it is remarkable that the same word should enter into numeral composition under such different meanings.

Nootka. [360]

10.	haiu.	
20.	tsakeits.	
30.	tsakeits ic haiu	= 20 + 10.
40.	atlek	= 2 × 20.
60.	katstsek	= 3 × 20.
80.	moyek	= 4 × 20.
100.	sutc'ek	= 5 × 20.
120.	nop'ok	= 6 × 20.
140.	atlpok	= 7 × 20.
160.	atlakutlek	= 8 × 20.
180.	ts'owakutlek	= 9 × 20.
200.	haiuk	= 10 × 20.

This scale is quinary-vigesimal, with no apparent decimal element in its composition. But the derivation of some of the terms used is detected with difficulty. In the following scale the vigesimal structure is still more obscure.

Tsimshian. [361]

10.	gy'ap.	
20.	kyedeel	= 1 man.
30.	gulewulgy'ap.	
40.	t'epqadalgyitk, or tqalpqwulgyap.	
50.	kctoncwulgyap.	
100.	kcenecal.	
200.	k'pal.	
300.	k'pal te kcenecal	= 200 + 100.
400.	kyedal.	
500.	kyedal te kcenecal	= 400 + 100.
600.	gulalegyitk.	
700.	gulalegyitk te kcenecal	= 600 + 100.
800.	tqalpqtalegyitk.	
900.	tqalpqtalegyitk te kcenecal	= 800 + 100.
1000.	k'pal.	

To the unobservant eye this scale would certainly appear to contain no more than a trace of the vigesimal in its structure. But Dr. Boas, who is one of the most careful and accurate of investigators, says in his comment on this system: "It will be seen at once that this system is quinary-vigesimal.... In 20 we find the word *gyat*, man. The hundreds are identical with the numerals used in counting men (see p. 87), and then the quinary-vigesimal system is most evident."

Rio Norte Indians. [362]

20.	taiguaco.	
30.	taiguaco co juyopamauj ajte	$= 20 + 2 \times 5.$
40.	taiguaco ajte	$= 20 \times 2.$
50.	taiguaco ajte co juyopamauj ajte	$= 20 \times 2 + 5 \times 2.$

Caribs of Essiquibo, Guiana

10.	oween-abatoro.	
20.	owee-carena	= 1 person.
40.	oko-carena	= 2 persons.

| 60. | oroowa-carena | | = 3 persons. |

Otomi

10.	ra-tta.	
20.	na-te.	
30.	na-te-m'a-ratta	= 20 + 10.
40.	yo-te	= 2 × 30.
50.	yote-m'a-ratta	= 2 × 20 + 10.
60.	hiu-te	= 3 × 20.
70.	hiute-m'a-ratta	= 3 × 20 + 10.
80.	gooho-rate	= 4 × 20.
90.	gooho-rate-m'a ratta	= 4 × 20 + 10.
100.	cytta-te	= 5 × 20,
	or nanthebe	= 1 × 100.

Maya, Yucatan. [363]

1.	hun.	
10.	lahun	= it is finished.
20.	hunkal	= a measure, or more correctly, a fastening together.
30.	lahucakal	= 40 − 10?
40.	cakal	= 2 × 20.
50.	lahuyoxkal	= 60 − 10.
60.	oxkal	= 3 × 20.
70.	lahucankal	= 80 − 10.
80.	cankal	= 4 × 20.
90.	lahuyokal	= 100 − 10.
100.	hokal	= 5 × 20.
110.	lahu uackal	= 120 − 10.
120.	uackal	= 6 × 20.
130.	lahu uuckal	= 140 − 10.
140.	uuckal	= 7 × 20.

200.   lahuncal   = 10 × 20.

300.   holhukal   = 15 × 20.

400.   hunbak   = 1 tying around.

500.   hotubak.

600.   lahutubak

800.   calbak   = 2 × 400.

900.   hotu yoxbak.

1000.   lahuyoxbak.

1200.   oxbak   = 3 × 400.

2000.   capic (modern).

8000.   hunpic   = 1 sack.

16,000.   ca pic (ancient).

160,000.   calab   = a filling full

3,200,000. kinchil.

64,000,000. hunalau.

In the Maya scale we have one of the best and most extended examples of vigesimal numeration ever developed by any race. To show in a more striking and forcible manner the perfect regularity of the system, the following tabulation is made of the various Maya units, which will correspond to the "10 units make one ten, 10 tens make one hundred, 10 hundreds make one thousand," etc., which old-fashioned arithmetic compelled us to learn in childhood. The scale is just as regular by twenties in Maya as by tens in English. It is

364

20 hun	= 1 kal	= 20.
20 kal	= 1 bak	= 400.
20 bak	= 1 pic	= 8000.
20 pic	= 1 calab	= 160,000.
20 calab	= 1{ kinchil / tzotzceh }	= 3,200,000.
20 kinchil	= 1 alau	= 64,000,000.

The original meaning of *pic*, given in the scale as "a sack," was rather "a short petticoat, somtimes used as a sack." The word *tzotzceh* signified "deerskin." No reason can be given for the choice of this word as a numeral, though the appropriateness of the others is sufficiently manifest. No evidence of digital numeration appears in the first 10 units, but, judging from the almost universal practice of the Indian tribes of both North and South America, such may readily have been the origin of Maya counting. Whatever its origin, it certainly expanded and grew into a system whose perfection challenges our admiration. It was worthy of the splendid civilization of this unfortunate race, and, through its simplicity and regularity, bears ample testimony to the intellectual capacity which originated it.

The only example of vigesimal reckoning which is comparable with that of the Mayas is the system employed by their northern neighbours, the Nahuatl, or, as they are more commonly designated, the Aztecs of Mexico. This system is quite as pure and quite as simple as the Maya, but differs from it in some important particulars. In its first 20 numerals it is quinary (see p. 141), and as a system must be regarded as quinary-vigesimal. The Maya scale is decimal through its first 20 numerals, and, if it is to be regarded as a mixed scale, must be characterized as decimal-vigesimal. But in both these instances the vigesimal element preponderates so strongly that these, in common with their kindred number systems of Mexico, Yucatan, and Central America, are always thought of and alluded to as vigesimal scales. On account of its importance, the Nahuatl system [365] is given in fuller detail than most of the other systems I have made use of.

10.	matlactli	= 2 hands.
20.	cempoalli	= 1 counting.
21.	cempoalli once	= 20-1.
22.	cempoalli omome	= 20-2.
30.	cempoalli ommatlactli	= 20-10.
31.	cempoalli ommatlactli once	= 20-10-1.
40.	ompoalli	= 2 × 20.
50.	ompoalli ommatlactli	= 40-10.
60.	eipoalli, or epoalli,	= 3 × 20.

70.	epoalli ommatlactli	= 60-10.
80.	nauhpoalli	= 4 × 20.
90.	nauhpoalli ommatlactli	= ~~90~~ 80-10.
100.	macuilpoalli	= 5 × 20.
120.	chiquacempoalli	= 6 × 20.
140.	chicompoalli	= 7 × 20.
160.	chicuepoalli	= 8 × 20.
180.	chiconauhpoalli	= 9 × 20.
200.	matlacpoalli	= 10 × 20.
220.	matlactli oncempoalli	= 11 × 20.
240.	matlactli omompoalli	= 12 × 20.
260.	matlactli omeipoalli	= 13 × 20.
280.	matlactli onnauhpoalli	= 14 × 20.
300.	caxtolpoalli	= 15 × 20.
320.	caxtolli oncempoalli.	
399.	caxtolli onnauhpoalli ipan caxtolli onnaui	= 19 × 20 + 19.
400.	centzontli	= 1 bunch of grass, or 1 tuft of hair.
800.	ometzontli	= 2 × 400.
1200.	eitzontli	= 3 × 400.
7600.	caxtolli onnauhtzontli	= 19 × 400.
8000.	cenxiquipilli, or cexiquipilli.	
160,000.	cempoalxiquipilli	= 20 × 8000.
3,200,000.	centzonxiquipilli	= 400 × 8000.
64,000,000.	cempoaltzonxiquipilli	= 20 × 400 × 8000.

Up to 160,000 the Nahuatl system is as simple and regular in its construction as the English. But at this point it fails in the formation of a new unit, or rather in the expression of its new unit by a simple word; and in the expression of all higher numbers it is forced to resort in some measure to compound terms, just as the English might have done had it not been able to borrow from the Italian.

The higher numeral terms, under such conditions, rapidly become complex and cumbersome, as the following analysis of the number 1,279,999,999 shows. [366] The analysis will be readily understood when it is remembered that *ipan* signifies plus. *Caxtolli onnauhpoaltzonxiquipilli ipan caxtolli onnauhtzonxiquipilli ipan caxtolli onnauhpoalxiquipilli ipan caxtolli onnauhxiquipilli ipan caxtolli onnauhtzontli ipan caxtolli onnauhpoalli ipan caxtolli onnaui; i.e.* 1,216,000,000 + 60,800,000 + 3,040,000 + 152,000 + 7600 + 380 + 19. To show the compounding which takes place in the higher numerals, the analysis may be made more literally, thus: $(15 + 4) \times 20 \times 400 \times 8000$ + (15 + 4) × 400 × 8000 + (15 + 4) × 20 × 8000 + (15 + 4) × 8000 + (15 + 4) × 400 + (15 + 4) × 20 + 15 + 4. Of course this resolution suffers from the fact that it is given in digits arranged in accordance with decimal notation, while the Nahuatl numerals express values by a base twice as great. This gives the effect of a complexity and awkwardness greater than really existed in the actual use of the scale. Except for the presence of the quinary element the number just given is really expressed with just as great simplicity as it could be in English words if our words "million" and "billion" were replaced by "thousand thousand" and "thousand thousand thousand." If Mexico had remained undisturbed by Europeans, and science and commerce had been left to their natural growth and development, uncompounded words would undoubtedly have been found for the higher units, 160,000, 3,200,000, etc., and the system thus rendered as simple as it is possible for a quinary-vigesimal system to be.

Other number scales of this region are given as follows:

Huasteca. [367]

10.	laluh.	
20.	hum-inic	= 1 man.
30.	hum-inic-lahu	= 1 man 10.
40.	tzab-inic	= 2 men.
50.	tzab-inic-lahu	= 2 men 10.
60.	ox-inic	= 3 men.
70.	ox-inic-lahu	= 3 men 10.
80.	tze-tnic	= 4 men.

90.	tze-ynic-kal-laluh	= 4 men and 10.
100.	bo-inic	= 5 men.
200.	tzab-bo-inic	= 2 × 5 men.
300.	ox-bo-inic	= 3 × 5 men.
400.	tsa-bo-inic	= 4 × 5 men.
600.	acac-bo-inic	= 6 × 5 men.
800.	huaxic-bo-inic	= 8 × 5 men.
1000.	xi.	
8000.	huaxic-xi	= 8-1000.

The essentially vigesimal character of this system changes in the formation of some of the higher numerals, and a suspicion of the decimal enters. One hundred is *boinic*, 5 men; but 200, instead of being simply *lahuh-inic*, 10 men, is *tsa-bo-inic*, 2 × 100, or more strictly, 2 times 5 men. Similarly, 300 is 3 × 100, 400 is 4 × 100, etc. The word for 1000 is simple instead of compound, and the thousands appear to be formed wholly on the decimal base. A comparison of this scale with that of the Nahuatl shows how much inferior it is to the latter, both in simplicity and consistency.

Totonaco. [368]

10.	cauh.	
20.	puxam.	
30.	puxamacauh	= 20 + 10.
40.	tipuxam	= 2 × 20.
50.	tipuxamacauh	= 40 + 10.
60.	totonpuxam	= 3 × 20.
100.	quitziz puxum	= 5 × 20.
200.	copuxam	= 10 × 20.
400.	tontaman.	
1000.	titamanacopuxam	= 2 × 400 + 200.

The essential character of the vigesimal element is shown by the last two numerals. *Tontamen*, the square of 20, is a simple word, and

1000 is, as it should be, 2 times 400, plus 200. It is most unfortunate that the numeral for 8000, the cube of 20, is not given.

Cora. [369]

10.	tamoamata.	
20.	cei-tevi.	
30.	ceitevi apoan tamoamata	$= 20 + 10.$
40.	huapoa-tevi	$= 2 \times 20.$
60.	huaeica-tevi	$= 3 \times 20.$
100.	anxu-tevi	$= 5 \times 20.$
400.	ceitevi-tevi	$= 20 \times 20.$

Closely allied with the Maya numerals and method of counting are those of the Quiches of Guatemala. The resemblance is so obvious that no detail in the Quiche scale calls for special mention.

Quiche. [370]

10.	lahuh.	
20.	hu-uinac	$= 1$ man.
30.	hu-uinac-lahuh	$= 20 + 10.$
40.	ca-uinac	$= 2$ men.
50.	lahu-r-ox-kal	$= -10 + 3 \times 20.$
60.	ox-kal	$= 3 \times 20.$
70.	lahu-u-humuch	$= -10 + 80.$
80.	humuch.	
90.	lahu-r-ho-kal	$= -10 + 100.$
100.	hokal.	
1000.	o-tuc-rox-o-kal.	

Among South American vigesimal systems, the best known is that of the Chibchas or Muyscas of the Bogota region, which was obtained at an early date by the missionaries who laboured among them. This system is much less extensive than that of some of the more northern races; but it is as extensive as almost any other South American system with the exception of the Peruvian, which was, however, a pure decimal system. As has already been stated, the

native races of South America were, as a rule, exceedingly deficient
in regard to the number sense. Their scales are rude, and show great
poverty, both in formation of numeral words and in the actual extent
to which counting was carried. If extended as far as 20, these
scales are likely to become vigesimal, but many stop far short of that
limit, and no inconsiderable number of them fail to reach even 5. In
this respect we are reminded of the Australian scales, which were so
rudimentary as really to preclude any proper use of the word "system"
in connection with them. Counting among the South American
tribes was often equally limited, and even less regular. Following
are the significant numerals of the scale in question:

Chibcha, or Muysca. [371]

10.	hubchibica.	
20.	quihica ubchihica	= thus says the foot, 10 = 10-10,
	or gueta	= house.
30.	guetas asaqui ubchihica	= 20 + 10.
40.	gue-bosa	= 20 × 2.
60.	gue-mica	= 20 × 3.
80.	gue-muyhica	= 20 × 4.
100.	gue-hisca	= 20 × 5.

Nagranda. [372]

10.	guha.	
20.	dino.	
30.	'badiñoguhanu	= 20 + 10.
40.	apudiño	= 2 × 20.
50.	apudiñoguhanu	= 2 × 20 + 10.
60.	asudiño	= 3 × 20.
70.	asudiñoguhanu	= 3 × 20 + 10.
80.	acudiño	= 4 × 20.
90.	acudiñoguhanu	= 4 × 20 + 10.
100.	huisudiño	= 5 × 20,
	or guhamba	= great 10.

200.	guahadiño	$= 10 \times 20$.
400.	diñoamba	= great 20.
1000.	guhaisudiño	$= 10 \times 5 \times 20$.
2000.	hisudiñoamba	= 5 great 20's.
4000.	guhadiñoamba	= 10 great 20's.

In considering the influence on the manners and customs of any people which could properly be ascribed to the use among them of any other base than 10, it must not be forgotten that no races, save those using that base, have ever attained any great degree of civilization, with the exception of the ancient Aztecs and their immediate neighbours, north and south. For reasons already pointed out, no highly civilized race has ever used an exclusively quinary system; and all that can be said of the influence of this mode of counting is that it gives rise to the habit of collecting objects in groups of five, rather than of ten, when any attempt is being made to ascertain their sum. In the case of the subsidiary base 12, for which the Teutonic races have always shown such a fondness, the dozen and gross of commerce, the divisions of English money, and of our common weights and measures are probably an outgrowth of this preference; and the Babylonian base, 60, has fastened upon the world forever a sexagesimal method of dividing time, and of measuring the circumference of the circle.

The advanced civilization attained by the races of Mexico and Central America render it possible to see some of the effects of vigesimal counting, just as a single thought will show how our entire lives are influenced by our habit of counting by tens. Among the Aztecs the universal unit was 20. A load of cloaks, of dresses, or other articles of convenient size, was 20. Time was divided into periods of 20 days each. The armies were numbered by divisions of 8000; [373] and in countless other ways the vigesimal element of numbers entered into their lives, just as the decimal enters into ours; and it is to be supposed that they found it as useful and as convenient for all measuring purposes as we find our own system; as the tradesman of to-day finds the duodecimal system of commerce; or as the Babylonians of old found that singularly curious system, the sexagesimal. Habituation, the laws which the habits and customs of every-day life impose upon us, are so powerful, that our instinctive readiness to make use of any concept depends, not on the intrinsic perfection or imperfection which pertains to it, but on the familiarity with which previous use has invested it. Hence, while one race may use a decimal, another a quinary-vigesimal, and another a sexagesimal scale, and while one system may actually be inherently superior to another, no user of one method of reckoning need ever think of any other method as possessing practical inconveniences, of which those employing it are ever conscious. And, to cite a single instance which illustrates the unconscious daily use of two modes of reckoning in one scale, we have only to think of the singular vigesimal fragment which remains to this day imbedded in the numeral scale of the French. In counting from 70 to 100, or in using any number which lies between those limits, no Frenchman is conscious of employing a method of numeration less simple or less convenient in any particular, than when he is at work with the strictly decimal portions of his scale. He passes from the one style of counting to the other, and from the second back to the first again, entirely unconscious of any break or change; entirely unconscious, in fact, that he is using any particular system, except that which the daily habit of years has made a part himself.

Deep regret must be felt by every student of philology, that the primitive meanings of simple numerals have been so generally lost. But, just as the pebble on the beach has been worn and rounded by the beating of the waves and by other pebbles, until no trace of its original form is left, and until we can say of it now only that it is quartz, or that it is diorite, so too the numerals of many languages have suffered from the attrition of the ages, until all semblance of their origin has been lost, and we can say of them only that they are numerals. Beyond a certain point we can carry the study neither of number nor of number words. At that point both the mathematician and the philologist must pause, and leave everything beyond to the speculations of those who delight in nothing else so much as in pure theory.

The End.

Footnotes:

[1] Brinton, D. G., *Essays of an Americanist*, p. 406; and *American Race*, p. 359.

[2] This information I received from Dr. Brinton by letter.

[3] Tylor, *Primitive Culture*, Vol. I. p. 240.

[4] *Nature*, Vol. XXXIII. p. 45.

[5] Spix and Martius, *Travels in Brazil*, Tr. from German by H. E. Lloyd, Vol. II. p. 255.

[6] De Flacourt, *Histoire de le grande Isle de Madagascar*, ch. xxviii. Quoted by Peacock, *Encyc. Met.*, Vol. I. p. 393.

[7] Bellamy, Elizabeth W., *Atlantic Monthly*, March, 1893, p. 317.

[8] *Grundriss der Sprachwissenschaft*, Bd. III. Abt. i., p. 94.

[9] Pruner-Bey, *Bulletin de la Société d'Anthr. de Paris*, 1861, p. 462.

[10] "Manual Concepts," *Am. Anthropologist*, 1892, p. 292.

[11] Tylor, *Primitive Culture*, Vol. I. p. 245.

[12] *Op. cit., loc. cit.*

[13] "Aboriginal Inhabitants of Andaman Islands," *Journ. Anth. Inst.*, 1882, p. 100.

[14] Morice, A., *Revue d'Anthropologie*, 1878, p. 634.

[15] Macdonald, J., "Manners, Customs, etc., of South African Tribes," *Journ. Anthr. Inst.*, 1889, p. 290. About a dozen tribes are enumerated by Mr. Macdonald: Pondos, Tembucs, Bacas, Tolas, etc.

[16] Codrington, R. H., *Melanesians, their Anthropology and Folk-Lore*, p. 353.

[17] *E.g.* the Zuñis. See Cushing's paper quoted above.

[18] Haddon, A. C., "Ethnography Western Tribes Torres Strait," *Journ. Anth. Inst.*, 1889, p. 305. For a similar method, see *Life in the Southern Isles*, by W. W. Gill.

[19] Tylor, *Primitive Culture*, Vol. I. p. 246.

[20] Brinton, D. G., Letter of Sept. 23, 1893.

²¹ *Ibid*. The reference for the Mbocobi, *infra*, is the same. See also Brinton's *American Race*, p. 361.

²² Tylor, *Primitive Culture*, Vol. I. p. 243.

²³ *Op. cit., loc. cit.*

²⁴ Hyades, *Bulletin de la Société d'Anthr. de Paris*, 1887, p. 340.

²⁵ Wiener, C., *Pérou et Bolivie*, p. 360.

²⁶ Marcoy, P., *Travels in South America*, Vol. II p. 47. According to the same authority, most of the tribes of the Upper Amazon cannot count above 2 or 3 except by reduplication.

²⁷ *Op. cit.*, Vol. II. p. 281.

²⁸ *Glossaria Linguarum Brasiliensium*. Bororos, p. 15; Guachi, p. 133; Carajas, p. 265.

²⁹ Curr, E. M., *The Australian Race*, Vol. I. p. 282. The next eight lists are, in order, from I. p. 294, III. p. 424, III. p. 114, III. p. 124, II. p. 344, II. p. 308, I. p. 314, III. p. 314, respectively.

³⁰ Bonwick, J., *The Daily Life and Origin of the Tasmanians*, p. 144.

³¹ Latham, *Comparative Philology*, p. 336.

³² *The Australian Race*, Vol. I. p. 205.

³³ Mackenzie, A., "Native Australian Langs.," *Journ. Anthr. Inst.*, 1874, p. 263.

³⁴ Curr, *The Australian Race*, Vol. II. p. 134. The next four lists are from II. p. 4, I. p. 322, I. p. 346, and I. p. 398, respectively.

³⁵ Curr, *op. cit.*, Vol. III. p. 50.

³⁶ *Op. cit.*, Vol. III. p. 236.

³⁷ Müller, *Sprachwissenschaft*. II. i. p. 23.

³⁸ *Op. cit.*, II. i. p. 31.

³⁹ Bonwick, *op. cit.*, p. 143.

⁴⁰ Curr, *op. cit.*, Vol. I. p. 31.

⁴¹ Deschamps, *L'Anthropologie*, 1891, p. 318.

⁴² Man, E. H. *Aboriginal Inhabitants of the Andaman Islands*, p. 32.

[43] Müller, *Sprachwissenschaft*, I. ii. p. 29.

[44] Oldfield, A., Tr. Eth. Soc. Vol. III. p. 291.

[45] Bancroft, H. H., *Native Races*, Vol. I. p. 564.

[46] "Notes on Counting, etc., among the Eskimos of Point Barrow." *Am. Anthrop.*, 1890, p. 38.

[47] *Second Voyage*, p. 556.

[48] *Personal Narrative*, Vol. I. p. 311.

[49] Burton, B. F., *Mem. Anthr. Soc. of London*, Vol. I. p. 314.

[50] *Confessions.* In collected works, Edinburgh, 1890, Vol. III. p. 337.

[51] Ellis, Robert, *On Numerals as Signs of Primeval Unity.* See also *Peruvia Scythia*, by the same author.

[52] Stanley, H. M., *In Darkest Africa*, Vol. II. p. 493.

[53] Stanley, H. M., *Through the Dark Continent*, Vol. II. p. 486.

[54] Haumontè, Parisot, Adam, *Grammaire et Vocabulaire de la Langue Taensa*, p. 20.

[55] Chamberlain, A. F., *Lang. of the Mississaga Indians of Skugog. Vocab.*

[56] Boas, Fr., *Sixth Report on the Indians of the Northwest*, p. 105.

[57] Beauregard, O., *Bulletin de la Soc. d'Anthr. de Paris*, 1886, p. 526.

[58] Ray, S. H., *Journ. Anthr. Inst.*, 1891, p. 8.

[59] *Op. cit.*, p. 12.

[60] Müller, *Sprachwissenschaft*, IV. i. p. 136.

[61] Brinton, *The Maya Chronicles*, p. 50.

[62] Trumbull, *On Numerals in Am. Ind. Lang.*, p. 35.

[63] Boas, Fr. This information was received directly from Dr. Boas. It has never before been published.

[64] Bancroft, H. H., *Native Races*, Vol. II. p. 753. See also p. 199, *infra*.

[65] Mann, A., "Notes on the Numeral Syst. of the Yoruba Nation," *Journ. Anth. Inst.*, 1886, p. 59, *et seq.*

[66] Müller, *Sprachwissenschaft*, IV. i. p. 202.

[67] Trumbull, J. H., *On Numerals in Am. Ind. Langs.*, p. 11.

[68] Cushing, F. H., "Manual Concepts," *Am. Anthr.*, 1892, p. 289.

[69] Grimm, *Geschichte der deutschen Sprache*, Vol. I. p. 239.

[70] Murdoch, J., *American Anthropologist*, 1890, p. 39.

[71] Kleinschmidt, S., *Grammatik der Grönlandischen Sprache*, p. 37.

[72] Brinton, *The Arawak Lang. of Guiana*, p. 4.

[73] Petitot, E., *Dictionnaire de la langue Dènè-Dindjie*, p. lv.

[74] Gilij, F. S., *Saggio di Storia Am.*, Vol. II. p. 333.

[75] Müller, *Sprachwissenschaft*, II. i. p. 389.

[76] *Op. cit.*, p. 395.

[77] Müller, *Sprachwissenschaft*, II. i. p. 438.

[78] Peacock, "Arithmetic," in *Encyc. Metropolitana*, 1, p. 480.

[79] Brinton, D. G., "The Betoya Dialects," *Proc. Am. Philos. Soc.*, 1892, p. 273.

[80] Ridley, W., "Report on Australian Languages and Traditions." *Journ. Anth. Inst.*, 1873, p. 262.

[81] Gatschet, "Gram. Klamath Lang." *U. S. Geog. and Geol. Survey*, Vol. II. part 1, pp. 524 and 536.

[82] Letter of Nov. 17, 1893.

[83] Müller, *Sprachwissenschaft*, II. i. p. 439.

[84] Hale, "Indians of No. West. Am.," *Tr. Am. Eth. Soc.*, Vol. II. p. 82.

[85] Brinton, D. G., *Studies in So. Am. Native Languages*, p. 25.

[86] *Tr. Am. Philological Association*, 1874, p. 41.

[87] Tylor, *Primitive Culture*, Vol. I. p. 251.

[88] Müller, *Sprachwissenschaft*, IV. i. p. 27.

[89] See *infra*, Chapter VII.

[90] Ellis, A. B., *Ewe Speaking Peoples*, etc., p. 253.

91 Tylor, *Primitive Culture*, Vol. I. p. 256.

92 Stanley, *In Darkest Africa*, Vol. II. p. 493.

93 Chamberlain, A. F., *Proc. Brit. Ass. Adv. of Sci.*, 1892, p. 599.

94 Boas, Fr., "Sixth Report on Northwestern Tribes of Canada," *Proc. Brit. Ass. Adv. Sci.*, 1890, p. 657.

95 Hale, H., "Indians of Northwestern Am.," *Tr. Am. Eth. Soc.*, Vol. II. p. 88.

96 *Op. cit.*, p. 95.

97 Müller, *Sprachwissenschaft*, II. ii. p. 147.

98 Schoolcraft, *Archives of Aboriginal Knowledge*, Vol. IV. p. 429.

99 Du Chaillu, P. B., *Tr. Eth. Soc.*, London, Vol. I. p. 315.

100 Latham, R. G., *Essays, chiefly Philological and Ethnographical*, p. 247. The above are so unlike anything else in the world, that they are not to be accepted without careful verification.

101 Pott, *Zählmethode*, p. 45.

102 Gatschet, A. S., *The Karankawa Indians, the Coast People of Texas.* The meanings of 6, 7, 8, and 9 are conjectural with me.

103 Stanley, H. M., *In Darkest Africa*, Vol. II. p. 492.

104 Müller, *Sprachwissenschaft*, II. i. p. 317.

105 Toy, C. H., *Trans. Am. Phil. Assn.*, 1878, p. 29.

106 Burton, R. F., *Mem. Anthrop. Soc. of London.* 1, p. 314. In the illustration which follows, Burton gives 6820, instead of 4820; which is obviously a misprint.

107 Dobrizhoffer, *History of the Abipones*, Vol. II. p. 169.

108 Sayce, A. H., *Comparative Philology*, p. 254.

109 *Tr. Eth. Society of London* , Vol. III. p. 291.

110 Ray, S. H., *Journ. Anthr. Inst.*, 1889, p. 501.

111 Stanley, *In Darkest Africa*, Vol. II. p. 492.

112 *Op. cit., loc. cit.*

113 Tylor, *Primitive Culture*, Vol. I. p. 249.

[114] Müller, *Sprachwissenschaft*, IV. i. p. 36.

[115] Martius, *Glos. Ling. Brasil.*, p. 271.

[116] Tylor, *Primitive Culture*, Vol. I. p. 248.

[117] Roth, H. Ling, *Aborigines of Tasmania*, p. 146.

[118] Lull, E. P., *Tr. Am. Phil, Soc.*, 1873, p. 108.

[119] Ray, S. H. "Sketch of Api Gram.," *Journ. Anthr. Inst.*, 1888, p. 300.

[120] Kleinschmidt, S., *Grammatik der Grönlandischen Spr.*, p. 39.

[121] Müller, *Sprachwissenschaft*, I. ii. p. 184.

[122] *Op. cit.*, I. ii. p. 18, and II. i. p. 222.

[123] Squier, G. E., *Nicaragua*, Vol. II. p. 326.

[124] Schoolcraft, H. R., *Archives of Aboriginal Knowledge*, Vol. II. p. 208.

[125] Tylor, *Primitive Culture*, Vol. I. p. 264.

[126] Goedel, "Ethnol. des Soussous," *Bull. de la Soc. d'Anthr. de Paris*, 1892, p. 185.

[127] Ellis, W., *History of Madagascar*, Vol. I. p. 507.

[128] Beauregard, O., *Bull. de la Soc. d'Anthr. de Paris*, 1886, p. 236.

[129] Schoolcraft, H. R., *Archives of Aboriginal Knowledge*, Vol. II. p. 207.

[130] Tylor, *Primitive Culture*, Vol. I. p. 249.

[131] *Op. cit.* Vol. I. p. 250.

[132] Peacock, *Encyc. Metropolitana*, 1, p. 478.

[133] *Op. cit., loc. cit.*

[134] Schoolcraft, H. R., *Archives of Aboriginal Knowledge*, Vol. II. p. 213.

[135] *Op. cit.*, p. 216.

[136] *Op. cit.*, p. 206.

[137] Mariner, *Gram. Tonga Lang.*, last part of book. [Not paged.]

[138] Morice, A. G., "The Déné Langs," *Trans. Can. Inst.*, March 1890, p. 186.

[139] Boas, Fr., "Fifth Report on the Northwestern Tribes of Canada," *Proc. Brit. Ass. Adv. of Science*, 1889, p. 881.

[140] *Do. Sixth Rep.*, 1890, pp. 684, 686, 687.

[141] *Op. cit.*, p. 658.

[142] Bancroft, H. H., *Native Races*, Vol. II. p. 499.

[143] *Tr. Ethnological Soc. of London*, Vol. IV. p. 92.

[144] Any Hebrew lexicon.

[145] Schröder, P., *Die Phönizische Sprache*, p. 184 *et seq.*

[146] Müller, *Sprachwissenschaft*, II. ii. p. 147.

[147] *On Numerals in Am. Indian Languages.*

[148] Ellis, A. B., *Ewe Speaking Peoples*, etc., p. 253. The meanings here given are partly conjectural.

[149] Pott, *Zählmethode*, p. 29.

[150] Schoolcraft, *op. cit.*, Vol. IV. p. 429.

[151] Trumbull, *op. cit.*

[152] Chamberlain, A. F., *Lang, of the Mississaga Indians*, Vocab.

[153] Crawfurd, *Hist. Ind. Archipelago*, 1, p. 258.

[154] Hale, H., *Eth. and Philol.*, Vol. VII.; Wilkes, *Expl. Expedition*, Phil. 1846, p. 172.

[155] Crawfurd, *op. cit.*, 1, p. 258.

[156] *Op. cit., loc. cit.*

[157] Bancroft, H. H., *Native Races*, Vol. II. p. 498.

[158] Vignoli, T., *Myth and Science*, p. 203.

[159] Codrington, R. H., *The Melanesian Languages*, p. 249.

[160] *Op. cit., loc. cit.*

[161] Codrington, R. H., *The Melanesian Languages*, p. 249.

[162] Wickersham, J., "Japanese Art on Puget Sound," *Am. Antiq.*, 1894, p. 79.

[163] Codrington, R. H., *op. cit.*, p. 250.

[164] Tylor, *Primitive Culture*, Vol. I. p. 252.

[165] Compare a similar table by Chase, *Proc. Amer. Philos. Soc.*, 1865, p. 23.

[166] *Leibnitzii Opera*, III. p. 346.

[167] Pruner-Bey, *Bulletin de la Soc. d'Anthr. de Paris*, 1860, p. 486.

[168] Curr, E. M., *The Australian Race*, Vol. I. p. 32.

[169] Haddon, A. C., "Western Tribes of the Torres Straits," *Journ. Anthr. Inst.*, 1889, p. 303.

[170] Taplin, Rev. G., "Notes on a Table of Australian Languages," *Journ. Anthr. Inst.*, 1872, p. 88. The first nine scales are taken from this source.

[171] Latham, R. G., *Comparative Philology*, p. 352.

[172] It will be observed that this list differs slightly from that given in Chapter II.

[173] Curr, E. M., *The Australian Race*, Vol. III. p. 684.

[174] Bonwick, *Tasmania*, p. 143.

[175] Lang, J. D., *Queensland*, p. 435.

[176] Bonwick, *Tasmania*, p. 143.

[177] Müller, *Sprachwissenschaft*, II. i. p. 58.

[178] *Op. cit.*, II. i. p. 70.

[179] *Op. cit.*, II. i. p. 23.

[180] Barlow, H., "Aboriginal Dialects of Queensland," *Journ. Anth. Inst.*, 1873, p. 171.

[181] Curr, E. M., *The Australian Race*, Vol. II. p. 26.

[182] *Op. cit.*, Vol. II. p. 208.

[183] *Op. cit.*, Vol. II. p. 278.

[184] *Op. cit.*, Vol. II. p. 288.

[185] *Op. cit.*, Vol. I. p. 258.

[186] *Op. cit.*, Vol. I. p. 316.

[187] *Op. cit.*, Vol. III. p. 32. The next ten lists are taken from the same volume, pp. 282, 288, 340, 376, 432, 506, 530, 558, 560, 588, respectively.

[188] Brinton, *The American Race*, p. 351.

[189] Martius, *Glossaria Ling. Brazil.*, p. 307.

[190] *Op. cit.*, p. 148.

[191] Müller, *Sprachwissenschaft*, II. i. p. 438.

[192] Peacock, "Arithmetic," *Encyc. Metropolitana*, 1, p. 480.

[193] Brinton, *Studies in So. Am. Native Langs.*, p. 67.

[194] *Op. cit., loc. cit.*

[195] Brinton, *Studies in So. Am. Native Langs.*, p. 67. The meanings of the numerals are from Peacock, *Encyc. Metropolitana*, 1, p. 480.

[196] Mason, *Journ. As. Soc. of Bengal*, Vol. XXVI. p. 146.

[197] Curr, E. M., *The Australian Race*, Vol. III. p. 108.

[198] Bancroft, H. H., *Native Races*, Vol. I. p. 274.

[199] Clarke, Hyde, *Journ. Anthr. Inst.*, 1872, p. clvii. In the article from which this is quoted, no evidence is given to substantiate the assertion made. It is to be received with great caution.

[200] Hale, H., *Wilkes Exploring Expedition*, Vol. VII. p. 172.

[201] *Op. cit.*, p. 248.

[202] Hale, *Ethnography and Philology*, p. 247.

[203] *Loc. cit.*

[204] Ellis, *Polynesian Researches*, Vol. IV. p. 341.

[205] Gill, W. W., *Myths and Songs of the South Pacific*, p. 325.

[206] Peacock, "Arithmetic," *Encyc. Metropolitana*, 1, p. 479.

[207] Peacock, *Encyc. Metropolitana*, 1, p. 480.

[208] *Sprachverschiedenheit*, p. 30.

209 Crawfurd, *History of the Indian Archipelago*, Vol. I. p. 256.

210 Pott, *Zählmethode*, p. 39.

211 *Op. cit.*, p. 41.

212 Müller, *Sprachwissenschaft*, II. i. p. 317. See also Chap. III., *supra*.

213 Long, S. H., *Expedition*, Vol. II. p. lxxviii.

214 Martius, *Glossaria Ling. Brasil.*, p. 246.

215 Hale, *Ethnography and Philology*, p. 434.

216 Müller, *Sprachwissenschaft*, II. ii. p. 82.

217 The information upon which the above statements are based was obtained from Mr. W. L. Williams, of Gisborne, N.Z.

218 *Primitive Culture*, Vol. I. p. 268.

219 Ralph, Julian, *Harper's Monthly*, Vol. 86, p. 184.

220 Lappenberg, J. M., *History of Eng. under the Anglo-Saxon Kings*, Vol. I. p. 82.

221 The compilation of this table was suggested by a comparison found in the *Bulletin Soc. Anth. de Paris*, 1886, p. 90.

222 Hale, *Ethnography and Philology*, p. 126.

223 Müller, *Sprachwissenschaft*, II. ii. p. 183.

224 Bachofen, J. J., *Antiquarische Briefe*, Vol. I. pp. 101–115, and Vol. II. pp. 1–90.

225 An extended table of this kind may be found in the last part of Nystrom's *Mechanics*.

226 Schubert, H., quoting Robert Flegel, in Neumayer's *Anleitung zu Wissenschaftlichen Beobachtung auf Reisen*, Vol. II. p. 290.

227 These numerals, and those in all the sets immediately following, except those for which the authority is given, are to be found in Chapter III.

228 Codrington, *The Melanesian Languages*, p. 222.

229 Müller, *Sprachwissenschaft*, II. ii. p. 83.

[230] *Op. cit.*, I. ii. p. 55. The next two are the same, p. 83 and p. 210. The meaning given for the Bari *puök* is wholly conjectural.

[231] Gallatin, "Semi-civilized Nations," *Tr. Am. Eth. Soc.*, Vol. I. p. 114.

[232] Müller, *Sprachwissenschaft*, II. ii. p. 80. Erromango, the same.

[233] Boas, Fr., *Proc. Brit. Ass'n. Adv. Science*, 1889, p. 857.

[234] Hankel, H., *Geschichte der Mathematik*, p. 20.

[235] Murdoch, J., "Eskimos of Point Barrow," *Am. Anthr.*, 1890, p. 40.

[236] Martius, *Glos. Ling. Brasil.*, p. 360.

[237] Du Graty, A. M., *La République du Paraguay*, p. 217.

[238] Codrington, *The Melanesian Languages*, p. 221.

[239] Müller, *Sprachwissenschaft*, II. i. p. 363.

[240] Spurrell, W., *Welsh Grammar*, p. 59.

[241] Olmos, André de, *Grammaire Nahuatl ou Mexicaine*, p. 191.

[242] Moncelon, *Bull. Soc. d'Anthr. de Paris*, 1885, p. 354. This is a purely digital scale, but unfortunately M. Moncelon does not give the meanings of any of the numerals except the last.

[243] Ellis, *Peruvia Scythia*, p. 37. Part of these numerals are from Martius, *Glos. Brasil.*, p. 210.

[244] Codrington, *The Melanesian Languages*, p. 236.

[245] Schweinfurth, G., *Linguistische Ergebnisse einer Reise nach Centralafrika*, p. 25.

[246] Park, M., *Travels in the Interior Districts of Africa*, p. 8.

[247] Pott, *Zählmethode*, p. 37.

[248] *Op. cit.*, p. 39.

[249] Müller, *Sprachwissenschaft*, IV. i. p. 101. The Kru scale, kindred with the Basa, is from the same page.

[250] Park, in Pinkerton's *Voyages and Travels*, Vol. XVI. p. 902.

[251] Park, *Travels*, Vol. I. p. 16.

252 Schweinfurth, G., *Linguistische Ergebnisse einer Reise nach Centralafrika*, p. 78.

253 Park, *Travels*, Vol. I. p. 58.

254 Goedel, "Ethnol. des Soussous," *Bull. Soc. Anth. Paris*, 1892, p. 185.

255 Müller, *Sprachwissenschaft*, I. ii. p. 114. The Temne scale is from the same page. These two languages are closely related.

256 *Op. cit.*, I. ii. p. 155.

257 *Op. cit.*, I. ii. p. 55.

258 Long, C. C., *Central Africa*, p. 330.

259 Müller, *Sprachwissenschaft*, IV. i. p. 105.

260 Pott, *Zählmethode*, p. 41.

261 Müller, *op. cit.*, I. ii. p. 140.

262 Müller, *Sprachwissenschaft*, IV. i. p. 81.

263 Pott, *Zählmethode*, p. 41.

264 Müller, *op. cit.*, I. ii., p. 210.

265 Pott, *Zählmethode*, p. 42.

266 Schweinfurth, *Linguistische Ergebnisse*, p. 59.

267 Müller, *Sprachwissenschaft*, I. ii. p. 261. The "ten" is not given.

268 Stanley, *Through the Dark Continent*, Vol. II. p. 490. Ki-Nyassa, the same page.

269 Müller, *op. cit.*, I. ii. p. 261.

270 Du Chaillu, *Adventures in Equatorial Africa*, p. 534.

271 Müller, *Sprachwissenschaft*, III. i. p. 65.

272 Du Chaillu, *Adventures in Equatorial Africa*, p. 533.

273 Müller, *op. cit.*, III. ii. p. 77.

274 Balbi, A., *L'Atlas Eth.*, Vol. I. p. 226. In Balbi's text 7 and 8 are ansposed. *Taru* for 5 is probably a misprint for *tana*.

275 Du Chaillu, *op. cit.*, p. 533. The next scale is *op. cit.*, p. 534.

[276] Beauregard, O., *Bull. Soc. Anth. de Paris*, 1886, p. 526.

[277] Pott, *Zählmethode*, p. 46.

[278] *Op. cit.*, p. 48.

[279] Turner, *Nineteen Years in Polynesia*, p. 536.

[280] Erskine, J. E., *Islands of the Western Pacific*, p. 341.

[281] *Op. cit.*, p. 400.

[282] Codrington, *Melanesian Languages*, pp. 235, 236.

[283] Peacock, *Encyc. Met.*, Vol. 1. p. 385. Peacock does not specify the dialect.

[284] Erskine, *Islands of the Western Pacific*, p. 360.

[285] Turner, G., *Samoa a Hundred Years Ago*, p. 373. The next three scales are from the same page of this work.

[286] Codrington, *Melanesian Languages*, p. 235. The next four scales are from the same page. Perhaps the meanings of the words for 6 to 9 are more properly "more 1," "more 2," etc. Codrington merely indicates their significations in a general way.

[287] Hale, *Ethnography and Philology*, p. 429. The meanings of 6 to 9 in this and the preceding are my conjectures.

[288] Müller, *Sprachwissenschaft*, IV. i. p. 124.

[289] Aymonier, E., *Dictionnaire Francaise-Cambodgien*.

[290] Müller, *Op. cit.*, II. i. p. 139.

[291] Müller, *Sprachwissenschaft*, II. i. p. 123.

[292] Wells, E. R., Jr., and John W. Kelly, Bureau of Ed., Circ. of Inf., No. 2, 1890.

[293] Pott, *Zählmethode*, p. 57.

[294] Müller, *Op. cit.*, II. i. p. 161.

[295] Petitot, *Vocabulaire Française Esquimau*, p. lv.

[296] Müller, *Sprachwissenschaft*, II. i. p. 253.

[297] Müller, *Op. cit.*, II. i. p. 179, and Kleinschmidt, *Grönlandisches Grammatik*.

[298] Adam, L., *Congres Int. des Am.*, 1877, p. 244 (see p. 162 *infra*).

[299] Gallatin, "Synopsis of Indian Tribes," *Trans. Am. Antq. Soc.*, 1836, p. 358. The next fourteen lists are, with the exception of the Micmac, from the same collection. The meanings are largely from Trumbull, *op. cit.*

[300] Schoolcraft, *Archives of Aboriginal Knowledge*, Vol. II. p. 211.

[301] Schoolcraft, *Archives of Aboriginal Knowledge*, Vol. V. p. 587.

[302] In the Dakota dialects 10 is expressed, as here, by a word signifying that the fingers, which have been bent down in counting, are now straightened out.

[303] Boas, *Fifth Report B. A. A. S.*, 1889. Reprint, p. 61.

[304] Boas, *Sixth Report B. A. A. S.*, 1890. Reprint, p. 117. Dr. Boas does not give the meanings assigned to 7 and 8, but merely states that they are derived from 2 and 3.

[305] *Op. cit.*, p. 117. The derivations for 6 and 7 are obvious, but the meanings are conjectural.

[306] Boas, *Sixth Report B. A. A. S.*, 1889. Reprint, pp. 158, 160. The meanings assigned to the Tsimshian 8 and to Bilqula 6 to 8 are conjectural.

[307] Hale, *Ethnography and Philology*, p. 619.

[308] *Op. cit., loc. cit.*

[309] Hale, *Ethnography and Philology*, p. 619.

[310] Müller, *Sprachwissenschaft*, II. i. p. 436.

[311] *Op. cit.*, IV. i. p. 167.

[312] *Op. cit.*, II. i. p. 282.

[313] *Op. cit.*, II. i. p. 287. The meanings given for the words for 7, 8, 9 are conjectures of my own.

[314] Müller, *Sprachwissenschaft*, II. i. p. 297.

[315] Pott, *Zählmethode*, p. 90.

[316] Müller, *op. cit.*, II. i. p. 379.

[317] Gallatin, "Semi-Civilized Nations of Mexico and Central America," *Tr. Am. Ethn. Soc.*, Vol. I. p. 114.

[318] Adam, Lucien, *Congres Internationale des Americanistes*, 1877, Vol. II. p. 244.

[319] Müller, *Sprachwissenschaft*, II. i. p. 395. I can only guess at the meanings of 6 to 9. They are obviously circumlocutions for 5-1, 5-2, etc.

[320] *Op. cit.*, p. 438. Müller has transposed these two scales. See Brinton's *Am. Race*, p. 358.

[321] Marcoy, P., *Tour du Monde*, 1866, 2ème sem. p. 148.

[322] *Op. cit.*, p. 132. The meanings are my own conjectures.

[323] An elaborate argument in support of this theory is to be found in Hervas' celebrated work, *Arithmetica di quasi tutte le nazioni conosciute.*

[324] See especially the lists of Hale, Gallatin, Trumbull, and Boas, to which references have been given above.

[325] Thiel, B. A., "Vocab. der Indianier in Costa Rica," *Archiv für Anth.*, xvi. p. 620.

[326] These three examples are from A. R. Wallace's *Narrative of Travels on the Amazon and Rio Negro*, vocab. Similar illustrations may be found in Martius' *Glos. Brasil.*

[327] Martius, *Glos. Brasil.*, p. 176.

[328] Adam, L., *Congres International des Americanistes*, 1877, Vol. II. p. 244. Given also *supra*, p. 53.

[329] O'Donovan, *Irish Grammar*, p. 123.

[330] Armstrong, R. A., *Gaelic Dict.*, p. xxi.

[331] Spurrell, *Welsh Dictionary*.

[332] Kelly, *Triglot Dict.*, pub. by the Manx Society.

[333] Guillome, J., *Grammaire Française-Bretonne*, p. 27.

[334] Gröber, G., *Grundriss der Romanischen Philologie*, Bd. I. p. 309.

[335] Pott, *Zählmethode*, p. 88.

[336] Van Eys, *Basque Grammar*, p. 27.

[337] Pott, *Zählmethode*, p. 101.

[338] *Op. cit.*, p. 78.

[339] Müller, *Sprachwissenschaft*, I. ii. p. 124.

[340] *Op. cit.*, p. 155.

[341] *Op. cit.*, p. 140.

[342] *Op. cit., loc. cit.*

[343] Schweinfurth, *Reise nach Centralafrika*, p. 25.

[344] Müller, *Sprachwissenschaft*, IV. i. p. 83.

[345] *Op. cit.*, IV. i. p. 81.

[346] *Op. cit.*, I. ii. p. 166.

[347] Long, C. C., *Central Africa*, p. 330.

[348] Peacock, *Encyc. Met.*, Vol. I. p. 388.

[349] Müller, *Sprachwissenschaft*, III. ii. p. 64. The next seven scales are from *op. cit.*, pp. 80, 137, 155, 182, 213.

[350] Pott, *Zählmethode*, p. 83.

[351] *Op. cit.*, p. 83,—Akari, p. 84; Circassia, p. 85.

[352] Müller, *Sprachwissenschaft*, II. i. p. 140.

[353] Pott, *Zählmethode*, p. 87.

[354] Müller, *Sprachwissenschaft*, II. ii. p. 346.

[355] *Op. cit.*, III. i. p. 130.

[356] Man, E. H., "Brief Account of the Nicobar Islands," *Journ. Anthr. Inst.*, 1885, p. 435.

[357] Wells, E. R., Jr., and Kelly, J. W., "Eng. Esk. and Esk. Eng. Vocab.," Bureau of Education Circular of Information, No. 2, 1890, p. 65.

[358] Petitot, E., *Vocabulaire Française Esquimau*, p. lv.

[359] Boas, Fr., *Proc. Brit. Ass. Adv. Sci.*, 1889, p. 857.

[360] Boas, *Sixth Report on the Northwestern Tribes of Canada*, p. 117.

361 Boas, Fr., *Fifth Report on the Northwestern Tribes of Canada*, p. 85.

362 Gallatin, *Semi-Civilized Nations*, p. 114. References for the next two are the same.

363 Bancroft, H. H., *Native Races of the Pacific States*, Vol. II. p. 763. The meanings are from Brinton's *Maya Chronicles*, p. 38 *et seq.*

364 Brinton, *Maya Chronicles*, p. 44.

365 Siméon Rémi, *Dictionnaire de la langue nahuatl*, p. xxxii.

366 An error occurs on p. xxxiv of the work from which these numerals are taken, which makes the number in question appear as 279,999,999 instead of 1,279,999,999.

367 Gallatin, "Semi-Civilized Nations of Mexico and Central America," *Tr. Am. Ethn. Soc.* Vol. I. p. 114.

368 Pott, *Zählmethode*, p. 89. The Totonacos were the first race Cortez encountered after landing in Mexico.

369 *Op. cit.*, p. 90. The Coras are of the Mexican state of Sonora.

370 Gallatin, *Semi-Civilized Nations*, p. 114.

371 Humboldt, *Recherches*, Vol. II. p. 112.

372 Squier, *Nicaragua*, Vol. II. p. 326.

373 Gallatin, *Semi-Civilized Nations*, p. 57.